Energy Efficiency Matters

About the series 'Books for the Concerned Citizen'

Leveraging the diverse expertise of its members in the subject domains and in publishing, the TERI Alumni Association is publishing a series of books on topics related to energy, resources, and the environment. The idea is to share information and, even more important, critical insights and understanding, with citizens who are keen to know more about some of the critical issues facing society and the world today but are lost in the deluge of information.

Our target audience is educated adults who are concerned about topical issues but lack the understanding to make sense of what they read or watch in the mass media—the series aims to equip them with conceptual tools and essential information not only to enrich their understanding but also to encourage them to act and thereby, albeit indirectly, further the UN Sustainable Development Goals.

The topics to be covered in the series and their respective subject-matter-specialist authors are listed below.

- **Rooftop solar***: Suneel Deambi and Shirish Garud
- **Coal***: Rakesh Kacker
- **Sustainable buildings***: Mili Majumdar and Minni Sastry
- **Nutraceuticals***: Mayurika Goel
- **Electricity***: Sanjeev S Ahluwalia
- **Clean transport***: Shri Prakash and Sharif Qamar
- **Climate change***: Manish Shrivastava
- **Energy efficiency**: Ajay Mathur and Leher V Thadani

already published and available for purchase

The publication of this series is financially supported by the Shakti Sustainable Energy Foundation. All books, printed and marketed by TERI, will be published latest by November 2022.

Energy Efficiency Matters

Ajay Mathur
Leher V Thadani

© TERI Alumni Association 2022
Second reprint 2023

ISBN: 978-93-94657-04-5

Suggested citation
Mathur A and Thadani L V. 2022. *Energy Efficiency Matters*. New Delhi: TERI Alumni Association. 74 pp.

TERI Alumni Association
Administrative Wing, TERI
Darbari Seth Block (ground floor)
Habitat Place
Lodhi Road
New Delhi – 110 003

Price Rs 299/-
For sales queries, please contact us at:
TERI Publications
The Energy and Resources Institute (TERI)
6C, Darbari Seth Block,
India Habitat Centre, Lodhi Road,
New Delhi 110 003, India
Tel: +91 011 2468 2100 or 7110 2100
Email: teripress@teri.res.in
Bookstore: https://bookstore.teri.res.in

For more information, contact
Ajay Mathur (amathur@cleanenergynet.in)
Leher V Thadani (lvthadani@gmail.com)

CONTENTS

Foreword	*vii*
Preface	*xi*
The concerned citizen's lexicon	1
Potential for energy efficiency	6
The household energy bill	14
Mindful energy consumption at home	21
Choosing and building energy-efficient homes	47
Energy efficiency beyond the home	56
Frequently asked questions	62
Conclusion	69
References	72

FOREWORD

I have known Ajay Mathur for over three decades, and have had the good fortune of partnering with him, and sharing platforms with him, throughout his leadership of several energy and sustainability-focused organizations in the public and private sectors, including during his previous role as Director General of TERI; Ajay played an especially critical and pioneering role in mainstreaming concerted policy action and informing public conversations to drive energy efficiency adoption during his tenure as Director General of Bureau of Energy Efficiency.

He is, therefore, uniquely positioned to advise individual and residential energy consumers on their respective roles and actions in achieving this much-needed energy transition. In this book, Ajay and Leher Thadani (who has a deep view on energy efficiency actions as a former public relations professional, who was deeply involved in promoting energy efficiency) have translated the abstract concepts and principles of responsible energy use that feature in policy, institutional and academic discourse, into tangible, practical daily practices and considerations that will have measurable impact on several fronts – reduced harmful emissions from a global standpoint, reduced energy import burden for the nation, increased electricity access across and within communities, and reduced energy bills for households and individuals.

Such a treatise is especially timely and necessary in the present environment. The world is facing a trifecta of crises – severe climate change events, energy shortages, and economic recessions - that can fundamentally alter human existence and the manner in which societies function; at the same time, we also find ourselves at the cusp of deploying clean energy technologies and programmes at a scale that can help avert negative impact in the long-term. Energy efficiency – the 'very first fuel' – is an essential component of all clean energy transitions, and urgent action is required to accelerate global efficiency progress. That is why I have always prioritised it in all our work at the International Energy Agency.

The awareness, attitudes and actions of end consumers towards energy use, especially at the individual and household levels, can have far-reaching repercussions across the global value chain, influencing international energy diplomacy and accordingly shifting the trajectory of legislative developments and institutional alignment that could see viable, necessary and meaningful climate action by all countries.

I therefore strongly urge all stakeholders across countries to act on the recommendations and practical guidelines detailed in the book that follows. While the chapters on the energy landscape and electricity bills may find more relevance with consumers in India, international audiences may realise the depth of the realities they share with their counterparts across borders and therefore develop a deeper appreciation for their priorities.

Of greatest relevance and resonance to all consumers across geographies will undoubtedly be the several chapters that detail the principles and practices for energy efficiency for residential consumers—while designing and selecting their homes, and while using energy conscientiously both within and outside their respective residences. Industry players and policy decision makers will also be well-advised to consider the recommendations in these chapters, and in the chapter that addresses consumers' common apprehensions, when designing products and policies that prioritise energy efficiency adoption.

True to his fashion in action and speech, Ajay, along with Leher Thadani, has authored a book that is engaging, accessible, and pertinent for readers of all ages and socio-cultural and economic backgrounds. I hope audiences will give it their fullest attention and consideration, which it merits, and wish it the greatest success.

Fatih Birol
Executive Director
International Energy Agency

November 2022

PREFACE

Energy is a key driver for the Indian economy and a prime enabler of diverse sectors including manufacturing, agriculture, health care, and railways. India not only continues to be a net importer, but its energy demand is also outpacing the growth of its GDP.

The resulting surge in energy prices for both commercial and household consumers in the country, along with ongoing efforts towards energy independence, indicates that the need to sustain, if not accelerate, the past momentum towards reducing energy intensity remains urgent and critical.

Currently accounting for nearly a quarter of the national energy demand, households are amongst the key players in the country's energy market and can significantly address this need from both immediate and long-term perspectives. Per capita electricity consumption has increased nearly 100-fold in the last 70 years, and with economic development, upward mobility, universal electrification, and the consequent greater access to technology and reliance on electrical appliances, residential consumption will accelerate. Efforts targeted at increasing awareness and adoption of energy efficiency measures amongst this group of consumers can therefore be expected to generate returns, both in the short and long terms, given the adaptability and heightened cost sensitivity of this consumer base.

The residential consumers' impact on the overarching efforts towards promoting widespread energy efficiency is undeniably promising. Over the last two decades, several organizations and the government have endeavoured to tap into this potential. Energy Efficiency Services Ltd has contributed to making select energy-efficiency appliances more affordable and to increasing their adoption by consumers, whereas BEE (Bureau of Energy Efficiency) has conducted programmes that have increased the awareness and appreciation of energy efficiency by schoolchildren through efforts such as the annual National Energy Conservation Day painting competition.

Even appliance manufacturers have moved quickly to incorporate energy-efficiency considerations into their market offerings over the last five years. However, the impact and the full potential of the newer technologies being developed will remain unrealized without a corresponding improvement in understanding and practice, and a change in mindset, on the part of consumers.

Surveys of consumer behaviour have indicated that, although residential consumers appreciate the need for energy efficiency, they lack clarity on its practice, achievement, and even their role and potential impact on the larger efforts towards both economic development and mitigation of climate change. This book attempts to provide consumers the information and clarity they need to become active and proactive stakeholders in the nationwide momentum towards energy conservation and efficiency.

In this book, we have endeavoured to empower the consumer, and to demonstrate and guide residential energy users on the ease and applicability of energy efficiency. We first attempt to establish a working understanding of India's energy landscape and ongoing efforts by institutions to propagate energy efficiency in industry and establish an enabling policy framework. With a view to enlighten consumers about how their electricity costs are calculated, we detail the various elements that constitute the electricity bill for residential consumers.

The primary focus of the book is to outline the principles of energy efficiency and highlight various practices that can be adopted across appliances, both at the time of purchase and during use, to reduce a household's electricity consumption and energy burden through interior and exterior design of homes. We also attempt to address some common questions and to align consumers' understanding of energy efficiency.

Throughout the book, we attempt to emphasize that all stakeholders have their own respective roles and responsibilities in facilitating the national and global energy transition, and that immediate and urgent action is incumbent upon each of them. Being residential energy consumers is a universal experience, and we therefore emphasize and indicate the role of this stakeholder group throughout the book.

We owe our gratitude to several individuals for their support and inputs throughout the development of this book.

We are deeply grateful to Dr Fatih Birol for penning the foreword, and for contributing his perspective on the larger energy landscape, which establishes the context in which this book was written. The many reports published by Dr Birol and the International Energy Agency have been immensely helpful in validating our assumptions, informing our understanding, and quantifying the immense progress that has been made – and yet needs to be made – towards achieving the necessary energy transition, which is a key element of effective and sustainable measures to achieve climate mitigation and resilience.

We received immense support from Mr Saurabh Kumar, former Managing Director, Energy Efficiency Services Ltd, who mobilized the national movement towards conscientious energy consumption and made technologies like LED bulbs, smart meters, and electric vehicles relevant, interesting, affordable, and accessible to households and communities across the nation and to its farthest reaches. The significance of his seminal work in developing and executing innovative business models, anchored in cross-stakeholder partnerships across the energy and technology value chains, to drive nationwide momentum for adoption of energy efficiency cannot be underlined enough, and we thank him for not only lending us his time and the benefit of his expertise and experience towards the improvement of this book, but also for his immeasurable contributions to national and global efforts towards climate mitigation.

We are also indebted to colleagues in the TERI Alumni Association and TERI Press for their invaluable support in making this book a reality: Mr Rakesh Kacker, former president of the TAA, for his ceaseless encouragement during the development of this book and for coordinating the multiple facets of the editing and publishing processes; Mr Yateendra Joshi and Mr P K Jayanthan for reviewing and editing the book for readability and coherence; and Ms Anupama Jauhary, Mr Rajiv Sharma, and Mr Vijay Nipane for developing engaging and comprehensible graphics that highlight the main features of the book and complement its narrative.

We thank and appreciate the tireless past, present, and future efforts of our colleagues and peers in the climate change, energy, sustainability, and associated realms in taking up the mantle of climate action, and for attempting all that is necessary, important, and urgent to cherish our collective home and preserve the ecosystems that will sustain livelihoods and enable the earth's inhabitants to enjoy a healthful future.

And so, we thank you, dear reader, for picking up this book, and in so doing, also accepting your role as a conscientious consumer and doing your part to advocate within your respective communities and to your leaders for more concerted action and innovation for a just and equitable use of the planet's finite resources.

Last, but in no measure the least, this book is a product of the collective effort of each member of our own respective families, households, and friends. Each of these individuals, who are far too numerous to name, created conducive conditions for authoring this manuscript and for embarking upon and completing this journey. We thank you, likely as you are to be unwittingly unaware of your contributions, for being all that you are. This book is inspired by you, written because of you, and dedicated to you.

The concerned citizen's lexicon

Avoided capacity generation represents gains from adopting energy efficiency or generating electricity from renewable sources of energy, and is usually expressed in either megawatts (MW) or million tonnes of oil equivalent (MTOE). The amount identifies the fossil-based electricity generation that was, or could be, avoided by using more efficient or renewable resources and also estimates the thermal energy that would have been used for similar activities during peak demand while also considering supply-side losses, constraints, and plant load factors. The metric is used in macro contexts to benchmark national progress towards mitigating climate change, and to demonstrate the cumulative market impact of energy-efficient appliances.

Carbon trading or **emissions trading** is a market-based system that enables countries and other entities, such as companies, to trade carbon credits as a commodity. Carbon credits are awarded to countries or companies for keeping their emissions of GHGs (greenhouse gases) within the limits assigned to them, expressed as CERs (certified emission rights) or AAUs (assigned amount units). Such credits can be sold for cash to countries or companies that need to reduce their emissions of GHGs either voluntarily or to comply with the relevant regulations. Originally envisaged in the Kyoto Protocol, the system was developed to incentivize the countries listed in Annex I of the UNFCCC, the United Nations Framework Convention on Climate Change, to transition to cleaner fuels and to develop climate-policy instruments that will eventually translate into lower global emissions on the one hand, while also investing in projects that reduce GHG emissions from the developing countries on the other.

Climate change refers to the ongoing long-term increase in the earth's surface and atmospheric temperature, and has been attributed primarily to the intensification of emissions due to human activity since the 19th century. Such activities range from power generation from fossil fuels to methane emissions from landfills. This temperature rise, and the subsequent volatility in weather patterns, have caused, and continue to effect, irreversible changes to the planet's ecological stability, which are now manifest as intense climate events, including droughts, forest fires, rising sea levels, flooding, melting of the polar ice caps, and declining biodiversity that adversely impact human life and livelihoods. The earth's temperature has already increased by 1.1 °C over the last 200 years, and scientists and experts across the world agree that a further increase by 0.5 °C will prove catastrophic, exacerbating world hunger, crop loss, water shortages, and threat to human life. Through international agreements like the Paris Agreement of 2015 and the Glasgow Climate Pact of 2021, institutions and nations across the world are endeavouring to identify methods and establish commitments to avert such a reality.

Decarbonization involves lowering the emissions of GHGs from industrial, transport, commercial, and residential sectors, and from other human activities. Applying to either specific industries or to whole sectors, decarbonization can refer to changes undertaken by an entity, be it a factory, company or government, to either change its sources of energy or to adapt more environment-friendly practices into its processes so that the quantum of emissions, also known as the **carbon footprint** of an activity, is decreased significantly as a proportion of the output of that activity. Decarbonization is a critical component of the efforts to lower emissions, with several governments even attempting to expand **carbon sinks**, which absorb carbon from the atmosphere and neutralize the impact of emissions, as part of their respective **national low-carbon strategies** and efforts to achieve a **low-carbon economy**.

Energy access is a household's connectivity and ready access to reliable and affordable electricity and clean energy for basic services and to meet the needs of a minimum standard of living, including clean cooking facilities. The term is also used interchangeably as a metric of a country's level of development and is calculated as a proportion of the number of people with ready access to electricity in their homes to the nation's total population.

Conversely, **energy poverty** describes the absence or lack of sufficient and affordable access to modern energy services and products through safe and reliable means. People and communities across the world, including those in the developed countries constituting 13% of the global population, suffer from energy poverty. Universal energy access, and ending energy poverty, by 2030 have thus been identified as Goal 7.1 in the **sustainable development goals** identified by the United Nations towards preserving the planet and improving the lives of its inhabitants.

Energy efficiency refers to efficient use of energy as a resource and entails using less energy to perform a task or complete an activity than would be consumed in the normal course. Energy efficiency also implies using energy conscientiously, thereby reducing wasteful use of energy, including minimizing **'vampire' power, or standby power**, which is continued electricity use by equipment and appliances even when they are switched off. Termed 'the first fuel' by the IEA (International Energy Agency), energy efficiency is an ongoing movement towards using energy optimally, either by modifying the technology and processes for achieving targeted outcomes or by lowering GHG-emitting activities to their minimal level through behavioural changes. Efficient use of energy is thus a vital component of mitigation and concerted climate action.

Energy intensity is the energy efficiency of a nation's economy, expressed as the ratio of its total energy consumption to its size as represented by the GDP. Although it may be tempting to deem a country whose energy intensity is lower to be more energy efficient, other factors need to be considered for a holistic evaluation and peer-group comparison, such as the level of overall economic development of the nation, access to quality and sustained power, reliance on fossil and renewable energy, and the rate of economic growth relative to the rate of household electrification.

Energy productivity, which is the inverse of energy intensity, measures economic output in relation to energy input. Usually measured as the ratio of purchasing power parity of a country's GDP to its primary energy use, a higher figure also reflects the robustness of the nation's infrastructure and indicates how effectively the infrastructure and resources are being used to serve the interests of the people and the objectives of economic growth. A key tenet of energy efficiency is thus to maximize energy productivity while reducing energy intensity.

Energy security refers to a nation's access to sufficient energy resources and supplies, as well as their equitable distribution, to sustain its population and economic activity at reasonable prices. India's Integrated Energy Policy: report of the expert committee, published in 2006, defined the country's energy security in terms of its ability to supply safe and convenient energy to all citizens to meet their requirements at affordable costs at all times, with a prescribed confidence level, while considering possible shocks and economic disruptions. As such, associations are often drawn with the wealth of a nation's natural resources. However, the ability to leverage and convert these resources for bolstering a country's economic strength and geopolitical prowess strongly influences this assessment, as well as its **energy independence**, which reveals a country's export dominance and self-reliance for its energy needs.

Energy transition is the ongoing effort to reduce dependence on fossil fuels for meeting the energy needs of countries and populations, and instead increasing adoption of renewable sources of energy, such as solar, wind, and hydropower as well as hydrogen. The transition entails achieving measurable shifts to renewable sources in each nation's energy mix, which captures the composition of sources for meeting a country's primary energy needs, and therefore necessitates structural changes to its energy infrastructure, while also shaping its decision-makers' approach to supply and demand. With direct implications for emissions, a nation's journey towards energy transition thus percolates every aspect of its economy and directly impacts the quality of life of its citizens, as well as global climate action, while also being affected by a country's technology and investment capabilities to accommodate transition needs and realities.

Greenhouse gas (GHG) emissions are gaseous emissions, by-products of energy use, that can absorb and capture infrared radiation, leading to heating of the atmosphere. These gases include carbon dioxide (CO_2), which is present in the highest concentration in the GHGs, as well as methane, nitrous oxide, and fluorinated gases, all of which collectively cause the greenhouse effect, which arises from the heat captured by these gases being reflected to the earth's surface, thereby contributing to climate change. Thus, the primary focus of mitigating climate change and safeguarding life on the planet is on encouraging nations to reduce their respective emissions intensity, the rate at which emissions are produced by human economic activity. Energy efficiency can accordingly have significant implications for moderating the energy intensity of growing economies.

Nationally determined contributions (NDCs) represent a nation's holistic plan to mitigate climate change by reducing its economy's net emissions while also adapting to the impacts of climate change. The commitments, or contributions, that nations make help experts to estimate and forecast the rate of global temperature rise, as well as the intensity and pace of the debilitating effects of climate change. Over 190 countries across the world delivered their respective NDCs for the landmark Paris Agreement, 2015, which provided a climate action framework for controlling global temperature rise. These contributions have since been updated by many countries.

Potential for energy efficiency

The broad national context

Since the beginning of the twenty-first century, India's economy has more than quadrupled in size, accompanied, and even powered, by a corresponding – although somewhat slower – growth in energy consumption, which is now triple of what it was in 2000. However, the vast majority of this energy supply continues to be heavily reliant on such fossil fuels as coal, lignite, gas, and diesel to the tune of nearly 60% of the country's electricity production. Even as cleaner renewable options such as hydro, wind, and solar have grown in India's electricity mix – to 39% according to the most recent figures (Ministry of Power 2022) – India's dependence on energy sources like coal is expected to continue for the foreseeable future to meet the country's development ambitions and its citizens' aspirations. Oil and natural gas have also continued to be the bases of India's economic development, with import dependence of nearly 80% in 2021.

These ambitions and aspirations nonetheless are juxtaposed with the increasingly threatening reality of accelerated climate change and severe climate events. Droughts and cyclones caused an economic loss of 87 billion dollars in India in 2020 alone (World Meteorological Organization 2021). India witnessed several catastrophic and fatal natural disasters in 2021, including the flash floods in Chamoli, Uttarakhand, that claimed 200 lives; the cyclones Tauktae, Yaas, Gulab, and Jawad; torrential landslides in Mahabaleshwar and Goa; and record high temperatures during extreme heat waves. The summer of 2022 was the hottest in central and north-western India in over 120 years, with temperatures rising by 2–4 °C in April compared to those in the past years (Kajal 2022).

Due to devastating human actions over the last two centuries, such occurrences have become more frequent – doubling in incidence since 2005, according to the climate vulnerability index of the CEEW (Council on Energy, Environment and Water) – and their human and economic impacts were greater than ever before. Such events are also expected to grow in scale and frequency if the current trends continue unchanged. In its report Climate change 2021: The physical science basis, the IPCC (Intergovernmental Panel on Climate Change) has projected that the South Asian subcontinent will see more intense and frequent heatwaves and stress from humid heat in the 21st century; this will translate into accelerated retreats of glaciers in the Himalayas, intense tropical cyclones and flooding, and erratic monsoons. Over 80% of India's population may be directly affected because it resides in areas exposed to extreme climate risks, with people in Andhra Pradesh, Assam, Bihar, Karnataka, and Maharashtra being increasingly vulnerable to floods, droughts, and cyclones (PTI 2021), and over 40% will suffer acute water scarcity by 2050 (Krishnan 2022).

India has initiated several measures to mitigate the impact of climate change. In 2008, the country adopted a formal plan, the NAPCC (National Action Plan on Climate Change), which continues to anchor its actions towards adapting to climate change and mitigating its adverse impacts. The plan encompasses eight national missions: the National Solar Mission to promote solar energy; the National Mission on Enhanced Energy Efficiency to focus on energy efficiency; the National Mission of Sustainable Habitat; the National Water Mission; the National Mission for Sustainable Agriculture; the National Mission for Sustaining the Himalayan Ecosystem for conserving natural resources, building resilience, and ensuring sustenance; the National Mission for a Green India to enhance ecosystem services; and the National Mission on Strategic Knowledge for Climate Change to strengthen research collaborations and technology development for climate action.

India has sought to balance economic growth and environmental protection in international conferences and dialogues by adopting and ratifying

- the 1987 Montreal Protocol that outlined the global phaseout of ozone depleting substances,
- the UNFCCC in 1993,
- the Kyoto Protocol in 2005, which bound countries to specific targets to lower their GHG emissions,
- the Nagoya Protocol in 2010 that outlined fair and equitable sharing of resources towards sustainable use of biodiversity,
- the Paris Agreement in 2015 that established the global action plan for restricting the global rise in temperatures to below 2 °C, and
- the Glasgow Climate Pact in 2021 that aims to strengthen global efforts and financing for building resilience to climate change and curbing the emissions of GHGs.

In accordance with its pledges in the Paris Agreement, India has committed to reduce the carbon emissions intensity of its GDP by 33%–35% and to increase, by 2030, the share of electricity generation capacity from sources other than fossil fuels by more than 40% of that in 2005. Further, in Glasgow, India also committed to achieve net-zero emissions by 2070. As part of this effort, by 2030 the Indian government has committed to meeting half its electricity requirements from renewable sources, thereby reducing the projected carbon emissions by 1 billion tonnes and also further reducing these emissions by another 2.5–3 billion tonnes by expanding forest and tree cover (Ministry of Environment, Forest and Climate Change 2021; Ministry of External Affairs 2021). However, it is important to note that India started reducing its energy dependence and managing the climate threat in a concerted manner long before all these measures, with the establishment of the Bureau of Energy Efficiency in 2002 under the aegis of the Energy Conservation Act. Aimed at reducing the energy and emissions intensity of India's economy, BEE anchors measures to promote adoption of energy efficiency across the country, including ideation and execution of the Standards and Labelling Programme that helps consumers understand the energy performance of common household electrical appliances through a star-rating system, development of ECBC (Energy Conservation Building Code) for commercial buildings, and enforcement of energy consumption norms in such energy-intensive industries as steel, textile, transport, cement, petrochemicals, and paper. The objective of all these programmes is to make energy-efficient technologies and practices the norm.

As part of this effort, along with Energy Efficiency Services Ltd, a sister concern under the Ministry of Power, BEE has also endeavoured to raise awareness and adoption of such energy-efficient technologies as LED bulbs through two more programmes, namely the 'Unnat Jyoti by Affordable LED for All' and the 'Street Lighting National Programme' at both household and industry levels.

These efforts have enabled energy consumers – at home, in offices and in industry – to adopt energy-efficient light bulbs, air conditioners, refrigerators, boilers, motors, and similar energy-saving items. Thus, although energy use has increased, the rate of increase in energy use has been slower than the rate of growth in the GDP.

The cumulative impact of such measures and foresight has been encouraging. India's energy intensity declined by more than 20% between 2005 and 2020 (Press Information Bureau 2020), while the emissions intensity decreased by 24% between 2005 and 2016 in line with the 2010 UNFCCC agreement (Ministry of Environment, Forest and Climate Change 2021).

That said, India's energy use and carbon emissions are expected to continue growing, even as the country endeavours to deliver on its commitments, made as part of the Paris Agreement, to enact policies and programmes that will help restrict global temperature rise to between 1.5 °C and 2 °C by 2050. India's energy sector is the largest contributor to its GHG emissions (approximately 75% in 2016), and the country is projected to nearly double its present installed capacity by 2030 and triple it by 2040, with renewable energy expected to constitute a growing portion of the added capacity. Electricity production accounted for nearly half of this sector's emissions in 2016; manufacturing and construction, nearly a fifth; and transport, over a tenth (Ministry of Environment, Forest and Climate Change 2021; International Energy Agency 2021).

It is therefore imperative to emphasize that the responsibility for effective mitigation that ensures a sustainable future for all cannot be borne by any one set of stakeholders alone, and that more robust and holistic action from the complete and collective cohort of stakeholders needs to be taken across sectors.

The role of residential consumers

Accounting for nearly a quarter of energy use, the potential contribution of the residential sector in lowering emissions, and indeed in reducing the impacts of climate change, is not insignificant, especially when one considers the pace at which residential consumption of electricity, petrol, diesel, and cooking gas has been increasing – due to the addition of new consumers, as well as increased consumption habits – thereby also indicating the scale on which demands on the country's energy ecosystem will grow. Following dedicated efforts towards universal electrification, 99% of the population and households across India had access to electricity in 2020, compared to 58.7% in 2000, as estimated by the World Bank. The growth on this metric for rural households is especially admirable, with access to electricity by rural population growing to 98.5% from merely 47.2%, while India's urban electrification rate grew to 100% from 88.8% in the same time frame (The World Bank 2022).

Households are thus amongst the key players in the country's energy market. By 2030, average household electricity use is expected to grow by 1.5–2.25 times that in 2015, an increase that translates to residential electricity consumption increasing by 5%–7.8% annually, with this segment of the economy expected to contribute up to a third of the country's overall demand (Ali 2018; Central Electricity Authority 2020; Spencer and Awasthy n.d.).

Since a large proportion of the country's electricity users are new, the need to educate them on responsible use and conservation of energy is more relevant, necessary, and urgent than ever before. This will help reduce energy demand and emissions of GHGs, thus achieving cost efficiencies, while also strengthening seamless and universal electricity supply to mitigate the adverse impacts of climate change. Technology and practices for energy conservation and efficiency can reduce electricity consumption of Indian households by up to 58%, a reduction that translates to an average saving of at least Rs 3900 in annual electricity costs, and also helps reduce 14 million tonnes of annual CO_2 emissions across the residential sector (Saleem et al. 2020).

Such savings from the residential sector alone could drive 10% of India's energy-saving potential (Mathur et al. 2018). As a whole, energy efficiency could reduce India's energy intensity by 19% by 2040 and contribute up to 56% to achieving India's 2030 target for lower emissions intensity. India improved its energy intensity by 20% between 2005 and 2020 by implementing the energy efficiency measures initiated by the government. These measures led to energy savings amounting to nearly 28 MTOE in 2019/20, reduced energy costs of Rs 1157 billion, and over 177 million tonnes in annual avoided CO_2 emissions (Alliance for Energy Efficient Economy 2021; Pimpalkhare 2020; Ministry of Environment, Forest and Climate Change 2021; International Energy Agency 2015).

Such energy savings help strengthen India's energy security, lower the country's dependence on imports, reduce overall costs in its development journey, while also improving the quality of life across the country (Figure 1). By 2040, with appropriate and necessary energy-efficiency interventions, India could reduce its import dependence on oil, coal, and natural gas by 18% and its carbon emissions by 19% (International Energy Agency 2015). These reductions, in turn, will also lower the concentrations of air

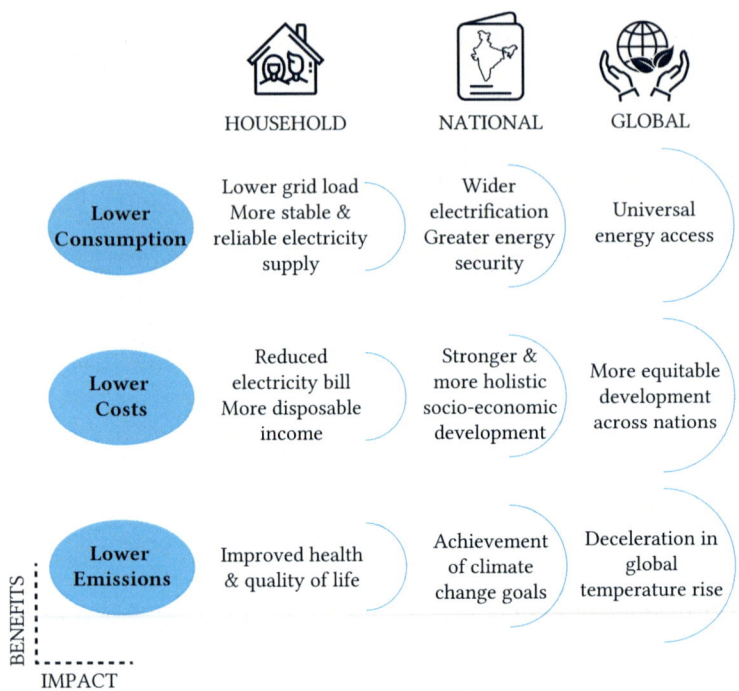

Figure 1 Benefits and impact of energy efficiency

pollutants, like fine particulate matter, sulfur oxide, and nitrogen oxide, which can reduce the average life expectancy in urban India by over 3 years.

Despite all its benefits, the vast majority of residential consumers in India are yet to tap into leveraging the full potential of 'the first fuel', as energy efficiency has been aptly termed by the IEA. Despite the ongoing awareness campaigns and endorsements, only a quarter of the population surveyed in the latest India Residential Energy Consumption Survey was found to be aware of the Standards and Labelling Programme and the BEE Star Label, the country's flagship programme for promoting energy-efficient technologies at the household level. The awareness in rural households is significantly lower than that in urban households. Although nearly all households use LED bulbs for their lighting needs, more than half of the surveyed households, on average, professed to have some knowledge of, or to have adopted, energy-efficient models of other common household appliances such as fans, television sets, hot-water geysers, and washing machines (Agrawal et al. 2020).

Furthermore, only 24% of the surveyed households cited energy saving as amongst the top two considerations in buying electrical appliances: the cost of the appliance, the popularity of its brand, and its average life carried more weight than its energy efficiency (Agrawal et al. 2020).

That said, even with only a small proportion of the population adopting energy-efficient technologies, the Standards and Labelling Programme saved approximately 56 billion units of electricity in 2020/21, translating to savings of Rs 300 billion and annual reductions in CO_2 emissions of approximately 46 million tonnes (Ministry of Power 2021). The implications of universal adoption of energy-efficient technologies are, therefore, monumental.

The chapters that follow attempt to bridge this gap in knowledge and awareness on part of households and residential consumers. The book seeks not only to educate residential consumers about using electricity and electrical appliances more responsibly and to convince such consumers of the merits of energy-efficient appliances, but also to guide consumers with key considerations to apply the principles of efficiency and conservation in broader contexts. The ambition is thus to not only help citizens unlock the full potential of energy-efficiency interventions, but also empower every member of India's citizenry with the agency and capabilities to act with intention towards mitigating the adverse effects of climate change.

The household energy bill

A fundamental motivation for conservation and responsible use of energy is the repercussion on household electricity costs. Electricity bills reveal how much electricity a household consumes, and that realization can result in individual, national, and energy-efficiency goals.

Electricity bills are generally issued once a month or once in two months by the power distribution company (discom, from now on) responsible for supplying electricity to consumers in a particular area. The electricity bill the customer receives at the end of every billing period reflects the costs incurred by the discom to supply electricity, which is a fixed component of the bill (the fixed cost), and the electricity consumed by the individual household, which is a variable component of the bill (the variable cost).

A household's electricity consumption is measured by a meter installed on the premises by the discom or the designated utility company. In the case of conventional meters, a representative of the discom visits the residence, reads the meter, and calculates the number of units consumed.

Since 2018, India has been adopting smart meters. A smart meter automatically records electricity consumption, voltage, current, and power factors and relays the data to the power supplier every 15 minutes, not only making manual readings redundant but also increasing accuracy and timeliness of billing. The meters have in-home displays enabling consumers to monitor energy usage in real time and are often supported by back-end utility systems that enable consumers to pay for electricity through either pre-billing or post-billing systems conveniently through consumer devices, including web and mobile applications.

Smart meters and advanced metering infrastructure that integrate 2-way digitally-powered communication and information management enable power companies to monitor power supply and reduce transmission and distribution (T&D) losses. Such monitoring, in turn, reduces the consumers' energy costs and enables consistent and widespread availability of electricity (Figure 2). Accordingly, the government aims to replace conventional meters with smart meters. Consumers in areas with high aggregate technical and commercial losses are to be served on priority for such a replacement programme in 2022.

The consumer is presented with a bill reflecting the following details.

Fixed charge A monthly charge predetermined by the respective state electricity regulatory commission for every discom and reviewed annually, the fixed charge reflects the costs to the discom for sourcing electricity from power plants and for installing and maintaining the materials and infrastructure for

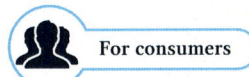

- Accurate and punctual electricity bills
- Real-time monitoring and control over electricity usage, especially during peak hours
- More information on electricity consumption, leading to more efficient consumption and smaller carbon footprint
- Convenient payment options
- Greater privacy, due to absence of manual meter readings
- Improved electricity supply

© Genus Power Infrastructure Ltd

  For distribution & generation companies

- Improved tracking of electricity supply and restoration following outages
- Reduced transmission and distribution losses
- Reduced operating and supply costs
- Improved energy efficiency that reduces peak load and need for additional power capacity

Figure 2 Advantages of smart meters

delivering electricity to consumers, including power lines and cables, poles, and electricity meters.

The fixed charge also shows the 'connected' or 'sanctioned' load, which is the maximum power that a household is allowed to use at any given time. This figure is fixed for an individual household at the time of applying for an electricity connection. Shown in the bill as kilowatts (kW), the connected load is determined by the discom and the consumer. This shows the anticipated load of a household on the electricity grid based on the total wattage of different electrical appliances and equipment used by the household as well as the size of the residential premises (in terms of floor space in square metres). Information on the appliances used by a household needs to be declared by consumers, and may be inspected by an engineer from the discom.

On occasion, discoms may also conduct and leverage the findings of 'load surveys' to establish the standards of living and the appliances used in a given area, subsequently adjusting the connected loads for households in that area; for instance, air conditioners reflect a major use of electricity in summer.

Discoms may increase the fixed charge, or levy additional charges, if a household exceeds its sanctioned power. This may feature on the bill as a demand charge.

The connected load also determines the type of supply and tariff connection – single-phase supply at low tension (LT) or multi-phase supply at high tension (HT) – granted to the household, which are usually reflected as the 'type of supply' and 'tariff category' on the bill. Whereas households with maximum connected loads of 3 kW can be serviced by single-phase connections of up to 220 volts, higher connected loads usually necessitate three-phase connections of up to 440 volts and therefore attract higher fixed costs.

Units consumed The electricity bill lists the units of electricity used by the consumer during the billing period as calculated from the meter reading. This consumption is due to the use of electrical equipment and appliances and thus represents the amount of electricity consumed by a household and the nature of electricity use by its members.

Although the electricity meter displays the cumulative units consumed, the number of units used during the billing period is calculated by subtracting the previous reading from the current reading. Multiplying this figure by the 'meter constant' or 'meter coefficient' provides the actual consumption in kilowatt hours (kWh). The meter constant, displayed on the meter as watt hours per pulse (Kh), is established by the meter's manufacturer.

The units consumed are applied to the 'tariff structure', also listed in detailed bills, thus yielding the 'energy charge' that constitutes the variable cost listed on a consumer's electricity bill. The tariff structure details the rates applicable at different consumption slabs, and is usually structured to incentivize low energy use, with the per-unit charges increasing with the number of units consumed.

It is important to note that the units consumed also impact the fixed charge. As mentioned earlier, consumption by a household higher than its sanctioned load puts strain on the grid, necessitating multi-phase connections that can accommodate the higher pressure.

Fuel surcharge Also known as the 'fuel cost adjustment charge', the fuel surcharge reflects changes in the cost of fuel to power generation companies, which they transfer to discoms. The discoms, in turn, pass the cost on to the consumer. Therefore, this charge is subject to national and global fuel prices and to consumer behaviour related to the use of electricity and energy use. The fuel surcharge is reviewed periodically – the frequency varying by state – by discoms and the state electricity regulatory commissions.

Electricity tax and electricity duty As with any consumable service, the consumer is taxed by the state government for supply of electricity, and that tax is yet another component of the electricity bill. This amount, paid by the consumer to the discom, is collected by the state government, with the rate of the tax varying as per each state's policy; some states levy the charge depending on all or part of the billed amount, whereas others charge the electricity duty as an additional cost on the units consumed. Each state's policy on this tax is described in its Electricity (Duty) Act. Consumers may review the act and stay apprised of the changes and developments in their respective states.

Rebates and incentives Discoms encourage energy-efficient and climate-conscious behaviour amongst consumers and ensure punctual payment of bills by way of a variety of offers that help households reduce their electricity bills.

Discoms offer rebates to those who opt for pre-paid metering, pay their bills before the due date, use electronic modes of payment, or improve their power factors. Discoms may also offer incentives to customers for adopting renewable sources of energy through devices like solar water heaters and rooftop solar systems.

Penalty Just as consumers are encouraged to use electricity judiciously, discoms also penalize customers if they draw more power than previously determined. Known as 'power factor' (PF), a low ratio drives heightened power generation and transmission, causing distribution losses, reducing efficient use of power, straining the transmission and distribution systems, and increasing GHG emissions.

Households using many appliances with motors, such as air conditioners and water pumps, which draw a large 'reactive' current in addition to the 'active' current, lowering the power factor of the electricity supply, may also be charged a PF penalty. The electricity bill therefore needs to be examined closely by the consumer to ensure accurate billing, as well as potential losses, leaks, or theft from the connection.

Sources of 'inductive loads' or 'reactive or phantom power' primarily drive low PF, and include inductive motors and transformers. Such equipment, used primarily in industrial or commercial complexes, usually operate on alternating current (AC) circuits, and both consume energy and also return it to the grid, thereby giving the impression that the equipment is consuming less energy than it actually does.

An efficient connection would register a PF close to 100%, whereas a connection with inductive loads would register significantly lower PF.

The real time PF can be checked from the electricity meter display. The PF penalty to the consumer is calculated based on the average PF for the billing cycle.

The bill also includes important information about the billing period, account number, and the type of electricity connection, which enables consumers to review and confirm that they have been billed correctly. The bill may also include other charges, namely arrears and penalties for pending payments, rent for electricity meters, as well as urban cess applied on behalf of municipal bodies and charges from other state utilities as stipulated in the established agreements between state government entities and discoms to leverage efficiencies in payment collection. The various elements of a typical electricity bill are shown in Figure 3.

Discoms generate on-the-spot as well as detailed bills, with both containing the information described above. On-the-spot bills are generated and provided to the consumer at the time of physical meter reading by an agent of the discom using a GPRS-enabled hand-held computing machine for efficient billing and collection, as well as for improved customer service. Detailed bills are issued by the discom after collecting the relevant details from the meter, either in person or remotely, and contain more details such as the type of connection, feeder codes, metering type and voltage, supply voltage, and ownership of the meter.

To sum up, the energy consumption habits of a household determine the variable cost of its electricity bill, specifically the energy charge and taxes or duties, and indirectly impact fixed

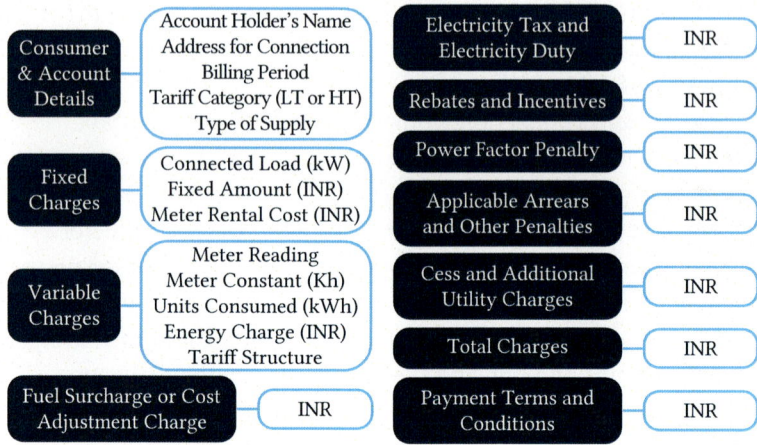

Figure 3 Elements of an electricity bill

charges. Excessive consumption may not only result in penalties and higher costs because of connected load to consumers in the short term, but also limit the availability and supply of energy and fuel, increasing fuel adjustment costs in the medium and long terms, as an increasing number of users connect to and scale up energy demands on the national grid.

Electricity bills thus contain a wealth of information to monitor consumption at micro and macro levels. The fundamentals of reading electricity bills can also be extended to reviewing and understanding bills generated for other forms of energy and resources that are typically used by consumers to meet their daily living needs and that are delivered through metered connections, such as piped gas for cooking, petrol or diesel at automotive refuelling stations, and even water. Tracking these costs and assessing consumption habits can thus empower consumers to extend the principles of efficiency to multiple aspects of their lives, enabling greater and wider conscientious use and conservation of resources.

Mindful energy consumption at home

Practising energy efficiency and conservation at the level of the individual consumer deliver significant benefits at the national and global levels: not only do households save energy and lower their energy bills, enhancing the quality and life of the appliances they use, but also assist the country in lowering the cost of importing energy and investments on transmission and distribution. Understanding how such costs are calculated, and examining the myriad ways in which consumers use energy, can help in delivering these benefits.

From powering fixtures and appliances in homes, various items of equipment in the workspace, facilities in commercial and retail buildings that deliver materials and services, electronic modes of passenger transport, and public infrastructure and amenities like roads that assist in economic development and citizen well-being, consumers use energy, specifically electricity, every moment of their day. Energy efficiency can be practised in nearly every facet, and thus is relevant to every aspect of a consumer's life.

Energy efficiency is practised based on two principles: (1) using existing consumer technology, equipment, and appliances in a responsible and conscientious manner and (2) regularly evaluating appliance needs and upgrading to their energy-efficient variants, when appropriate and necessary. The next section outlines these principles and describes the corresponding practices for most widely used appliances in Indian households. Although the practices described identify opportunities for energy efficiency and conservation primarily in homes, these principles are relevant to other areas as well. An energy-conscious user at home, we believe, can influence and lower energy use at work as well.

It is important to note that, although intent is the cornerstone of careful consumption, the benefits of energy efficiency can be availed of with minimal effort and adjustments in daily lives. The principles and practices described below demonstrate that being energy efficient is easy, convenient, and achievable at only marginal additional costs to the consumer; these are also aimed at changing the way energy is used from a broader perspective and context.

Principles of energy efficiency

Figure 4 summarizes the principles of energy efficiency, which are rephrased in this section as Rules 1 and 2, encompassing eight and three specific instructions respectively.

> **RULE 1** Use existing consumer technology, equipment, and appliances responsibly and conscientiously.

a/ Use what you need, when you need it, and only as much as you need. In today's fast-paced life that necessitates multi-tasking,

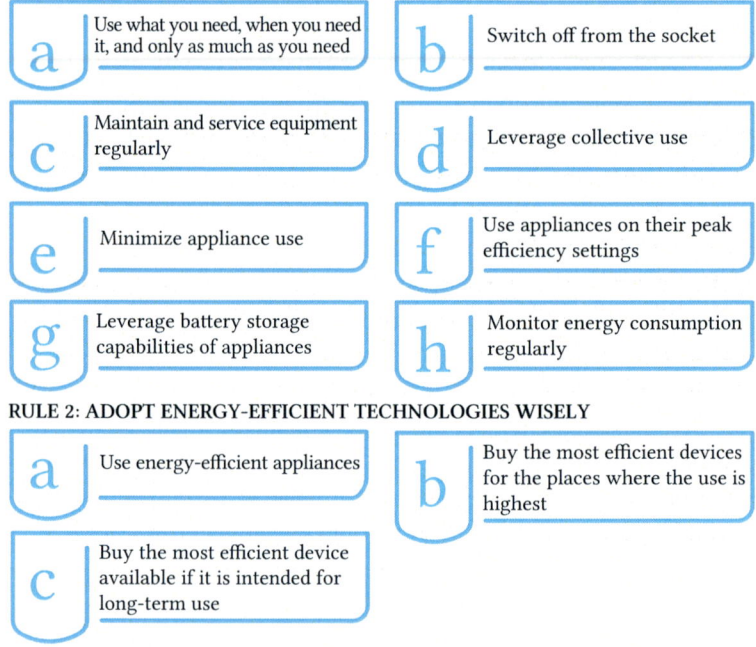

Figure 4 Principles of energy efficiency

consumers use multiple electrical appliances and equipment, often simultaneously. They may turn on multiple devices and appliances in anticipation of using them, but do not really use them all at the same time. Users must activate only those appliances and only at the time or in the sequence they are needed. For example, ovens and stovetops may be turned on only when food needs to be cooked or heated, and not at the beginning of the food preparation process. Similarly, fans and air conditioners may be switched on only in rooms and spaces that are occupied, and turned off in vacant or vacated areas.

Furthermore, it is also important for consumers to use energy and equipment only to the extent necessary. This would translate to not turning the gas flame up to high heat when food can be cooked on medium or low heat, setting the iron to the temperature setting appropriate for the clothes being ironed, and establishing the lowest fan speed, highest temperature settings for air conditioners, and lowest temperature settings for space heaters, that are comfortable for occupants. Most appliances introduced in the market over the last few years have been equipped with automatic turn-off/turn-on functions that help consumers regulate their use, and thus use energy and the appliance efficiently.

b/ *Switch off from the socket.* Usually when consumers finish using an appliance, they switch off the power button on the face or the body of the appliance—appliances that are turned off thus remain plugged into the sockets with the switch on and continue to consume electricity. This is especially true of appliances like television sets, digital video recorders (DVRs), coffee machines, and ovens that are operated with remote controls or are pre-programmable. The power button on these appliances does not cut off electricity supply, but merely transitions them into sleep or standby mode, which continues to consume energy, as these devices have to be ready to be reactivated automatically for the scheduled function or following a signal from the remote.

This latent energy use, often referred to as 'vampire power', can contribute up to a fifth of the monthly electricity usage. It is thus important that the devices must either be disconnected from the plug or switch that connects them to the rest of the circuit serving the residence. Consumers may re-plug or turn the wall switch on when they need to use the device again.

Unplugging can also help extend the life of equipment and appliances, protecting their internal circuitry from fluctuations in current due to frequent power outages and frequency variations across the country.

c/ *Maintain and service equipment regularly.* Through regular use, or long periods of non-use, the internal motor and parts of an appliance suffer wear and tear, resulting in malfunction or functioning at capacity that is less than optimal, thereby extending the duration of use and increasing the amount of energy consumed. It is therefore important for consumers to maintain and service the appliances regularly or get their peak and optimal functionality checked prior to use after an extended period of idleness or dormancy. For example, cleaning the dust filters of air conditioners at the beginning of a hot season can save 5%–10% of energy use over the season. Servicing also addresses any recalibrations that may be necessary to ensure that the equipment or machinery is capable of operating at peak efficiency and capacity.

Regular maintenance and servicing are recommended to extend the life of appliances like washing machines, air conditioners, refrigerators, and reverse-osmosis water filtration and supply systems. Consumers should also consult professionals about replacing or servicing of specific parts.

d/ *Leverage collective use.* As part of daily life, consumers are likely to use common facilities used by other people also simultaneously or sequentially. To increase efficiency, and to utilize the full potential of such facilities, consumers may systematize their use to minimize repetitive activation of an appliance. Examples of such facilities include scheduled public transport, carpooling by those travelling to the same location or in the same direction, encouraging groups and families to eat together to reduce heating and re-heating of food, and delaying the use of washing machines until a full load of laundry is collected.

The additional energy used by appliances by collating the tasks will be minimal compared to when an appliance is used multiple times to deliver the same benefit to fewer items or beneficiaries. Appropriate planning and collation of activities thus optimize the use of the appliances, contributing to their longevity, and reducing the energy and time taken for the task.

Over time, this practice of identifying opportunities for optimization can become habitual, thus helping consumers to use resources more efficiently across other aspects of their lives.

e/ *Minimize appliance use.* Electronic appliances and machinery have undoubtedly delivered convenience and saved time. With continued innovation and technology development, the appliances are able to perform increasingly complex tasks. That said, many tasks, especially in small volumes, can be carried out more efficiently and effectively without using any electronic appliances at all.

Consumers may thus evaluate the capacity of appliances, and the need to use them, beforehand to ensure that the advantages outweigh the costs and energy used for operation. An electronic food processor may be used for chopping vegetables or other ingredients in bulk, whereas a significantly smaller quantity may be diced with a manual chopper or a knife.

Accordingly, consumers may consider a supplementary device that will render the same results while consuming less electricity. For example, using an energy-efficient induction cooktop instead of gas stovetop for cooking and heating, or using 7-watt LED bulbs instead of 60-watt incandescent bulbs or 14-watt CFLs (compact fluorescence lamps) to illuminate only those areas that are being actively used for working, studying, or other space-specific activities.

f/ *Use appliances on their peak efficiency settings.* To derive the most value and benefit from a device, consumers must review the instructions issued by the manufacturer to understand how to use it at optimum efficiency. For example, air conditioners should be set at temperatures between 24 °C and 27.5°C to cool closed spaces in a sustained and energy-efficient manner. Similarly, cars in India may maintain the tyre pressure of at least 33 psi (pounds per square inch), or approximately 228 kPa or 2.3 kg/cm², for maximum fuel efficiency.

Establishing these settings and conditions may also necessitate resetting the devices' default factory settings, as the latter may correspond to test conditions set by the manufacturer or mandatory regulations, which may differ from the daily-use, real-world conditions in which the appliance is used. Hot-water geysers, for example, are set at 60 °C, but most people use water at only 40 °C for a comfortable bath. Changing this factory setting could reduce annual carbon emissions by nearly 200 kg per household and annual electricity bills by more than Rs 1200 (Ministry of Environment, Forest and Climate Change 2020). The user manuals may even provide an overview of such test conditions, but will definitely guide consumers on resetting the device to their needs and preferences.

g/ *Leverage battery storage capabilities of appliances.* Several present-day electronic devices are equipped with in-built rechargeable batteries or battery packs. Appliances like laptops, mobile phones, Bluetooth speakers, and electronic vehicles are a small subset. These appliances are designed to function fully for extended periods on their (rechargeable) batteries, but can also operate when connected to the main circuit.

To maximize battery capabilities and to use energy efficiently in a broader context, it may be beneficial to use such devices on their batteries, charging them only when the batteries are nearly discharged. This will deliver a more equitable use of electricity and also extend the life of the battery. When fully charged batteries are connected to a power outlet, they are used simply as carriers of electric current, and not for their storage and delivery capabilities.

An extension of this principle for larger equipment would be to install capacitors, which can help improve the power factor, especially in stand-alone residences that rely on motors to draw water from supply lines. To reduce strain on the residence's electrical network, the capacitor should be located close to the motor, switched on only when the motor is in use and switched off with the motor. An automatic power-factor controller may also be installed to minimize this operation.

As with any other equipment, the motor and accompanying capacitor should be used only when needed, and not left idling, which can cause the capacitor to either discharge or malfunction.

h/ *Monitor energy consumption regularly.* The impact of energy-efficiency practices on a household's consumption and bills can be realized within the short term, and often in a single billing cycle. Comparing the electricity bills across billing cycles before and after adopting energy-efficiency practices can reinforce their benefits, thus sustaining them over the medium and long terms, which can then transform them into habits that deliver maximum benefits to consumers.

Identifying key energy-efficiency activities that contribute to lowering electricity bills can thus reinforce appropriate behaviour, while also demonstrating energy-efficiency practices that actually work.

For households with smart meters, discoms also provide mobile applications that help to monitor and analyse consumption. Further technological developments are expected to deliver increasingly accurate and instant results, enabling consumers to track and understand electricity use reliably in real time.

RULE 2: Adopt energy-efficient technologies wisely
(Figure 4)

a/ *Use energy-efficient appliances.* In addition to using existing devices efficiently, consumers may also opt for new appliances designed to function efficiently and effectively while consuming minimal energy. Over the last few years, the consumer appliance industry has moved increasingly towards introducing energy-efficient models of common appliances, ranging from fans, TV sets, and lights to heavier equipment like air conditioners, refrigerators, and washing machines. Due to increasing awareness amongst consumers, the demand for such models has grown, making them price-competitive with their less efficient counterparts in the market. Simultaneously, Indian industry has responded to changing consumer preferences and built domestic capabilities to manufacture energy-efficient appliances, matching consumer demand with locally-built supply, thereby participating in the market transformation while strengthening its role in, and impact on, the global climate movement.

Furthermore, energy-efficient appliances save electricity; converted to reduction in electricity bills, the savings mean that these devices will pay for themselves in the medium to long terms. For example, each energy-efficient refrigerator and air conditioner can reduce annual electricity bills by Rs 700–1400 and each energy-efficient fan, by nearly Rs 350 (Ministry of Environment, Forest and Climate Change 2020). The acquisition cost of the equipment typically constitutes 20% or less of the total ownership cost: the remainder is the running cost over the equipment's lifetime. The higher upfront cost of energy-efficient equipment is offset by the savings due to lower operating costs, with energy-efficient equipment also lasting longer than its less efficient variants. Such savings, when compounded over energy-efficient equipment's usable lifetime, can help consumers offset a significant portion of their upfront purchase costs.

Much of this increased sensitization and market transformation has been constantly encouraged by the government over the last decade, with these initiatives continuing to be relevant today. The most pertinent, recognizable, and visible, as well as long-standing, example of such initiatives is BEE's Star Labelling Programme. Initiated in 2006, the programme informs consumers about the efficiency of various appliances, rating them on a scale of 1 (least efficient) to 5 (most efficient) and expressed in the form of stars (Figure 5), their number corresponding to the rating, as determined by independent quality assurance and testing agencies. Fixed prominently on the face of the appliance, the label also includes its power consumption to advise consumers on its lifetime running cost, as well as its energy efficiency ratio (EER), which, as the name indicates, quantifies its output or functionality against the electricity it consumes.

BEE also issues endorsement labels (Figure 6) to specific products, which certify them as the most efficient in their respective categories.

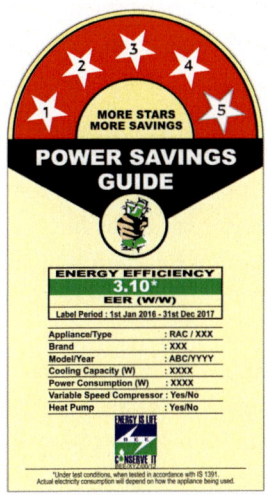

Figure 5 Sample BEE star label

Figure 6 Sample BEE endorsement label

These labels help consumers to purchase or upgrade to a new model of a given appliance. In its absence, consumers could request the retailer for information on a device's electricity consumption. At present, manufacturers of 10 categories of appliances sold in retail are required to include the label, while the label is optional or voluntary for 16 other categories (Table 1). However, the list of appliances covered by the BEE star label programme continues to expand, and more categories are added based on the patterns of consumer demand and market trends.

b/ *Buy the most efficient device for places where the use is highest.* While planning for and choosing new electronic appliances and devices, consumers may want to focus specifically on areas and functions that are used the most in a household. For example, fans or air conditioners in bedrooms run for at least 8 hours each day in summer, but will likely be used for far fewer hours and with lesser intensity in the dining room if this room is not also used for multiple purposes such as for working or studying. Thus, consumers may opt to install a 5-star fan or air conditioner in a multipurpose area or an area where the device will be used consistently for long periods of time, such as a children's play and study area or bedroom, and a 3-star fan or air conditioner in the dining room.

Similarly, lights for well-frequented areas, such as home entrances, that are lit up during most of the evening hours, may be prioritized for energy-efficiency upgrades over areas like bathrooms, where the lights are switched on only when that space is being used. It therefore makes sense for lights at entrances to be the most energy efficient.

c/ *Buy the most efficient device available if it is intended for long-term use.* The BEE star labels mentioned earlier are also revised and updated periodically, approximately every two years, in line with the programme's objective to reduce the energy bills of consumers and to motivate the development, manufacture, and adoption of energy-efficient products. Industries, and indeed the

Table 1 Appliances covered by BEE labelling programme

Mandatory	Voluntary
Frost-free refrigerator	Induction motors
Fluorescent tubes	Agricultural pump sets
Room air conditioners	Ceiling fans
Cassette, floor standing tower (ceiling, corner)	Domestic liquefied petroleum gas (LPG) stoves
Distribution transformers	Washing machines (front loading, drum type)
Direct-cool refrigerator	Washing machines (top loading and semi-automatic)
Stationary storage type electric water heaters (geysers)	Computers (notebook/laptop)
Colour television sets	Ballast (electronic/magnetic)
Variable-capacity air conditioners	Office equipment (printers, copiers, scanners)
LED lamps	Solid-state inverter
	Microwave ovens
	Diesel pump sets
	Diesel generators
	Chillers (air cooled and water cooled)
	Solar water heaters
	Light commercial air conditioners
	Deep freezers

SOURCE Bureau of Energy Efficiency

entire consumer technology and appliances market, have taken note of this effort and are constantly developing increasingly efficient technologies. The minimum benchmark for efficiency is thus revised, and a product with a 3-star efficiency rating today may not be as efficient as a 3-star product sold after two years;

consequently, today's 3-star product may have its label revised to 2 stars. Given the constant innovation in industry on this metric, consumers may not only want to buy the most efficient product available in the market to ensure that they are maximizing cost and functional efficiency benefits for the longest duration possible, but also routinely check the performance of their appliances and upgrade to more efficient models as needed and as budgets permit.

While using such recommended energy-efficient devices, consumers should continue to stay energy conscious. Although the operating cost of these more efficient devices is comparatively low, these too should be used conscientiously, applying the principles of energy efficiency, so that consumers derive the full benefits from the equipment.

Consumers may also consider using appliances powered by solar photovoltaic panels or similar forms of renewable energy, such as solar water heaters, solar ovens, and solar inverters. Each of these devices reduces annual carbon emissions by 500–1000 kg and annual electricity costs by nearly Rs 5000.

Solar-powered equipment also enjoys the incentives and subsidies offered by the central government and state governments.

Lastly, these devices require minimal maintenance and, in India's largely sunlit geography, can be used virtually throughout the year and most effectively during peak-load hours. If combined with battery storage, which is yet to become commercially viable, such devices can be used even when the sun is not shining, such as during the night or in the monsoon season when the sky is overcast for prolonged periods.

Energy efficiency practices for different categories of appliances

Energy-conscious practices for some of the most commonly used appliances across the country are described below.

Light bulbs
Use efficiently

- Use floor or free-standing lamps and table lamps if only a part of a room is being used or occupied instead of ceiling lights that illuminate the whole room.
- Clean bulbs, fluorescent tubes, and lampshades regularly to ensure effective functioning.
- Avoid switching on lights in rooms with natural light during daytime.
- Install light-coloured and loosely-woven curtains on windows and doors to take advantage of natural light during daytime.
- Turn off lights in unoccupied rooms.

Key considerations while buying efficient light bulbs
- Install LED (light-emitting diode) variants of bulbs and fluorescent tubes instead of conventional ones, as the former deliver the same lumen intensity as incandescent bulbs do at a fraction of the energy used and have a significantly longer life. Such LED bulbs are available in a variety of lighting temperatures that give white light, yellow light, and multicoloured light.
- Purchase products that bear the ISI mark (certified by the Bureau of Indian Standards) and the BEE star labels for quality assurance.
- Where possible, install motion-activated light panels that keep rooms lit as long as they are occupied, and that switch off automatically when vacated.

Fans
Use efficiently
- Turn off fans in unoccupied rooms and spaces.
- Clean the fan blades and the area around the rotor regularly to ensure effective use of the motor.
- Keep doors open to allow for cross-breeze across rooms if fans are running in multiple spaces simultaneously.

- For pedestal or tower fans, position the fans opposite to, or across the occupied area or area where the directed airflow is required. Fans may also be positioned across the room and facing an open window on cooler days to enable faster and uniform cooling across the room or space.
- Wear lighter fabric to moderate body temperature to reduce dependence on fans for comfort.
- Minimize the use of heat-radiating appliances like overhead lights and lamps during the day to lower the room temperature.

Key considerations while buying efficient fans
- Purchase BEE-star-labelled fans.
- Buy super-efficient fans that provide the same level of performance as 70-watt fans but use only half the energy.
- Use light-weight brushless direct current (BLDC) motors that minimize contact and, hence, rotor loss.
- Ensure that blade sizes and variable rotation speeds are suited to the size of the room in which the fan is being used to ensure adequate circulation.
- Install smart switches or auto-turn-off features that automatically turn the fan off after a set duration.

Air conditioners
Use efficiently

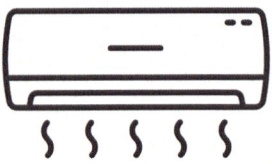

- Shut doors and windows to enclose the space that needs to be cooled to shorten the duration of running the air conditioner to achieve the desired comfort level.
- Set air conditioners at temperatures between 24 °C and 27.5 °C for optimal energy use and cooling performance.
- Use fans simultaneously with air conditioners set at higher temperatures (24 °C and above). This can deliver the same comfort as using an air conditioner set to a lower temperature but consumes less energy.

- Place air conditioners in the room in such a way that air is circulated throughout the room and not confined to nooks or niches that restrict airflow.
- Service air conditioners at least once a year to ensure high performance.
- Turn off air conditioners in unoccupied rooms and areas.
- Service air conditioners regularly, as instructed by the manufacturer and recommended by professional technician; clean and replace the air filters, as required, as part of the servicing.
- Keep inverters in shaded areas and with some space around them to improve the efficiency of the condenser and the evaporator in cooling and ejecting hot air.

Key considerations while buying efficient air conditioners
- Purchase BEE-star-labelled devices.
- Install inverter-based or split air conditioners, which track the external temperature and therefore always operate at peak efficiency. The impact on consumers' energy bills is nearly instantaneous, with the devices paying back for themselves in 3–4 years.
- Install air conditioners equipped with automatic temperature cut-off to monitor room temperature and accordingly regulate energy usage. This may also feature as electronically-controlled variable-speed compressors that adjust compressor speed to cooling needs.
- Determine the size of the air conditioner based on the area required to be cooled. Air conditioners with capacity (expressed in BTUs, or British thermal units, per hour) unaligned with the space they will cool will operate less effectively and efficiently than those of the appropriate capacity. A unit larger than necessary will cool the air before it can reduce the humidity in the room, causing rooms to be uncomfortably clammy, whereas units smaller than necessary will take significantly longer to bring the room to the desired temperature. Technology experts and salespeople can advise on the most suitable AC capacity based on the area of the

room, ceiling height, exposure to direct sunlight, average number of occupants at any given time, and location within the room.
- Buy units that use refrigerants with a lower global warming potential such as R-32 (HFC-32), R-454B, and R-290, to minimize harmful hydrofluorocarbon emissions that accentuate ozone depletion and accelerate climate change.

Space heaters
Use efficiently

- Shut doors and windows to enclose the space that needs to be heated to shorten the duration of running the heater to achieve the desired comfort level. Portable space heaters are most effective and energy efficient when used in small spaces or in single rooms, whereas central HVAC (heating, ventilation, and air conditioning) systems may prove more efficient and effective for larger spaces.
- Use the heater at the lowest temperature setting and for the shortest amount of time possible to achieve the desired comfort level. Heaters with in-built fans can be run at lower temperature settings, consuming less wattage and electricity while circulating air and warming the room more effectively. Also consider wearing warm clothes to reduce the use of the heater.
- Switch on the heater to warm rooms prior to sleeping, but do not run it unattended during the night to avoid unnecessary use and fire hazards.
- To minimize the risks of malfunction, position the heater on a flat and stable surface at least 60 cm (2 feet) away from water sources, flammable materials, and paths of movement; connect the heater directly to sockets instead of using extension cords or power strips.

Key considerations while buying efficient room heaters

- Assess the size of the room to be heated before selecting a suitable model, as the wattage or power output of a heater should directly correspond to room size.

- Buy thermostat-controlled fan heaters, or space heaters with multiple temperature or wattage and fan speed settings to enable the heater settings to be adjusted to individual comfort levels or preferences.
- Fix a safety switch that automatically turns off the heater if it is moved, shaken, or once the set temperature level is reached.
- Use long power cords to allow flexibility in the positioning of the heater in spots and areas they will serve the most.

Hot-water geysers
Use efficiently

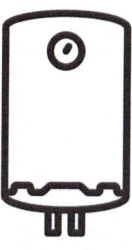

- Change the setting of the water temperature that suits individual comfort and tolerance. Geysers are by default set at 60 °C, consuming significantly more energy than average and are likely to provide water far hotter than needed.
- If hot water is needed by several people sequentially or for several sequential tasks, it may be used in quick succession to reduce the duration over which the geyser needs to stay switched on.
- Check the temperature of the water before switching on the geyser to determine its need. If the geyser is to be used, turn it off as soon as the water reaches the desired temperature. Several models are also sold with a thermostat or an automatic timer function that turns the geyser off once the desired water temperature is reached or after a pre-determined duration.
- Place the geyser in a shaded or covered area and closest to the point of use to minimize heat loss.
- Drain the geyser's tank at least once or twice a year to ensure cleanliness and safety of water as well as to remove any residue on heating elements for efficient functioning.
- Use low-flow faucets, fixtures, and taps, including water-saving showerheads, to reduce the pressure as well as the rate at which hot water is delivered.

Key considerations while buying efficient hot-water geysers

- Purchase BEE-star-labelled device, or a solar water heater.
- Consider the household's needs for hot water in terms of volume and usage patterns to determine the suitable capacity. If hot water is used by many in quick succession, a larger-capacity geyser may be appropriate; if used intermittently or in smaller volumes, a smaller-capacity geyser may prove beneficial and will also be energy efficient.
- Install appropriate insulating material or note any in-built features to reduce heat loss, as these can reduce the time for keeping a geyser switched on, retaining heat and keeping the water hot even when switched off, especially if the geyser is used in quick succession.
- Consider tankless geysers, which are more energy efficient than small-tank geysers, as they heat only as much water as is being consumed at any given moment.
- Install a temperature control, timer, or an automatic cut-off feature to regulate active use and energy consumption.
- Connect the geyser to solar panels, as viable and appropriate.

Television sets
Use efficiently

- Switch off the television set and all associated equipment using separate power sockets, including set-top boxes and speaker systems, from the socket when not in use.
- Lower the brightness of the screen to reduce the amount of energy consumed. TV sets are also equipped with an eco- or movie picture mode that automatically adjusts the brightness for ambient light for optimal viewing experience.
- Turn off the TV monitor, or lower its brightness, if only the audio is being used, for example while listening only to music or using only its audio features.

- Explore sleep settings on the television set to ensure that it switches off automatically after a stipulated time of inactivity by the viewer.

Key considerations while buying efficient television sets
- Purchase BEE-star-labelled devices.
- Buy LED or LCD (liquid crystal display) TV sets that include such features as backlighting and auto-brightness that reduce eyestrain and also deliver improved picture quality.
- Invest in smart power strips that can shut down appliances in standby mode to reduce vampire or 'phantom' load.
- Compare the room size or viewing distance to ensure that the size of the screen is optimal for the best experience while balancing energy consumption. An unduly large TV screen can overwhelm a room while consuming significant amount of electricity, and a smaller screen can strain viewers' eyes and prevent a comfortable viewing experience.

Refrigerators
Use efficiently
- Regularly defrost refrigerators and freezers.
- Keep at least 10 cm (4 inches) of space between the rear panel of the refrigerator and the wall to allow air circulation.
- Place the refrigerator away from heat sources, including direct sunlight, ovens, and stovetops.
- Set the temperature of the refrigerator to 2–4 °C or above, and that of the freezer to at least –16 °C for the refrigerator to operate at optimum efficiency. Also consult the user manual and manufacturer's directions to identify the most energy-efficient settings.
- Ensure that the refrigerator doors are shut properly and sealed airtight to prevent loss of cool air.

- In case the refrigerator is empty, place water bags in the freezer to ensure that cool air continues to be circulated; this may prove useful during power cuts or extended load shedding.
- Allow hot foods to cool to room temperature before placing them in the refrigerator.
- Cover liquids kept in the refrigerator to control internal moisture levels, thereby sustaining condenser performance.
- Avoid keeping the refrigerator door open for extended periods of time.
- Ensure that the vents inside the refrigerator are kept clear to facilitate effective air circulation throughout the refrigerator.
- Do not store items on top of the refrigerator. Ensure that adequate space is available for heat to escape, thereby assisting compressor function.
- Regularly clean and service the refrigerator as advised by the manufacturer or the technician. Pay special attention to dust collecting on condenser coils and air intake grills, which can strain motor function.

Key considerations while buying efficient refrigerators

- Purchase BEE-star-labelled devices.
- Buy refrigerator of capacity (expressed in litres) appropriate to its use after estimating the approximate weight and number of items to be stored. Proportionately smaller refrigerators will need to function more intensely than is efficient and effective, using unnecessarily more energy, whereas models that are larger than necessary entail higher upfront cost and lower use of their capacity, resulting in less effective use of electricity.
- Use glass shelves, which are better conductors of heat than plastic, keeping items cool from the bottom while also helping maintain a uniform temperature throughout the refrigerator.
- Ensure that the refrigerator door closes snuggly and tight, level with the refrigerator frame to prevent air leakage. Test the seal by wedging a single sheet of paper between the seal and the frame, if

the sheet stays in place and cannot be pulled away easily, the door is closed tight.
- Sign a maintenance and servicing contract directly with the manufacturer or with a local company authorized by the manufacturer.

Washing machines
Use efficiently

- Use cold-water cycles as much as possible to minimize geyser use unless specific loads of clothes mandate washing in hot water. A monthly maintenance wash with hot water and zero load is recommended to descale the machine and remove any bacteria.
- Use washing machines only when a full load is collected, and run wash cycles with as many clothes and fabrics that require similar washing treatments to optimize utility and maximize efficiency. A full load should ideally fill up to three-quarters of the washing drum to ensure uniform cleaning and washing of all the clothes.
- Pre-soak stained clothes, using safe stain removers to make the washing cycle more effective and to minimize the number of times the fabric needs to be washed, thereby making it last longer.
- Use the detergent appropriate for the washing machine and in quantities recommended by the manufacturer or prescribed in the user manual. The composition of the detergent or soap used for hand-washing laundry may be different and unsuitable for machine washing.
- Avoid running the washing machine during the hours of peak load. The discom can identify and advise on these hours, and may also inform residents of any time-of-use plans applicable in the area that can reduce the electricity costs of large and energy-hungry appliances, such as washing machines.
- Rinse the clothes and dry them outdoors to avoid the use of dryer machines. Washing machines may also be equipped with

a high-speed or extended spin cycle, which can help drain excess water from laundry prior to drying.
- Clean the washing machine's filter regularly to ensure smooth and effective working.

Key considerations while buying efficient washing machines
- Purchase BEE-star-labelled devices.
- Decide on the volume (measured in litres) of the washing machine appropriate for use, based on the estimated weight of a typical full load and frequency of use.
- Use the eco mode or the speed-perfect mode that moderates energy usage during the washing cycle.
- Front-loading machines use less water and are generally more resource conservative than top-loading machines, especially if hot water is being used for washing, although they take longer than top-loading machines to achieve the desired results. Top-loading machines use more water, which means the geyser has to work more, although the washing cycle itself may be shorter.

Stovetops
Use efficiently

- Prepare all the ingredients for cooking and keep within reach before turning on the gas or switching on the stove.
- Use pressure cookers to the extent possible to cook dishes with gravy or vegetables and lentils that take longer to soften, as pressure cookers conduct heat within the ingredients more effectively than other vessels can.
- If a pressure cooker is unsuitable, use broad- or flat-bottomed pots to cook food, because such pots present a larger surface in contact with the heat source and therefore use energy more effectively.
- As far as possible, cover pots and pans with lids while cooking

to allow equitable heat distribution and circulation of hot air, reducing cooking time.
- Use burners that are matched to the pot: use small burners for small pots and large burners for large pots.
- Pre-soak rice, beans, and lentils, which take longer to cook, to soften and reduce the time and energy required to convert them into consumable and edible dishes.
- Use the least amount of water necessary for preparing foods to reduce time and energy for the water to reach boiling or desired temperature.
- Reduce the flame once food starts boiling or cooking and at the earliest opportunity possible, turn off the gas to use the residual heat in the cooking utensil and food to complete the last stage of cooking. Electric stoves also retain heat for several minutes after they are switched off.
- Household members may eat together and as soon as possible after meal preparation to avoid reheating food.
- Minimize the number of times food is reheated; heat only as much as will be consumed at a time. Also use the smallest container appropriate for the quantity being heated to minimize the time and energy taken to heat both the container and the food.
- Before heating refrigerated foods, allow them to reach room temperature to reduce the time needed for them to reach the desired temperature for consumption.
- For reheating foods, use microwaves, which use half the energy and time consumed by conventional electric and gas stoves.
- Clean stovetop burners regularly to ensure their effective functioning.
- Ensure the flame on gas stovetops is blue at all times. If the flame is yellow or orange, consult a technician to check if the stovetop needs servicing or if the gas supply has leaks or impurities. Consult a technician if soot deposits not associated with burnt food are noticed on cookware or burners.

Key considerations while buying efficient stovetops
Gas stoves
- Purchase stoves approved by ISI or BIS.
- Opt for electric ignition models instead of those with a continually-burning or standing pilot.
- Install an exhaust or heat-exchanging range hood along with the gas stove to maintain indoor air quality and reduce pollutants emitted during cooking, such as nitrogen dioxide or carbon monoxide.

Induction cookstoves
- Purchase BEE-star-labelled devices.
- Opt for a device that can be powered by renewable energy sources.
- Purchase a device with programmable features that automate cooking time and temperature, including auto shut-off.
- Purchase appropriately sized ferromagnetic cookware along with the induction stove to ensure effective functioning. Alternatively, to allow for flexibility, zoneless units may also be considered.
- Ensure the device has sensors to measure the size of the cookware and control the temperature automatically, as well as to alert users to technical malfunctions.
- Ensure that the element has copper wire certified as pure.
- Consider power rating and wattage to match the likely intensity of use.
- Ensure provision of maintenance and warranty contracts.

As has been described above, energy-efficiency practices and technologies can deliver significant short- and long-term monetary benefits to individual consumers and households (Figure 7a and 7b). Along with reducing variable electricity usage, energy efficiency also impacts the power factor that influences the fixed costs in a household's electricity bill. Accordingly, consumers can request the discom to re-evaluate the connected or sanctioned load based on actual consumption patterns after adopting energy-efficient technologies and methods.

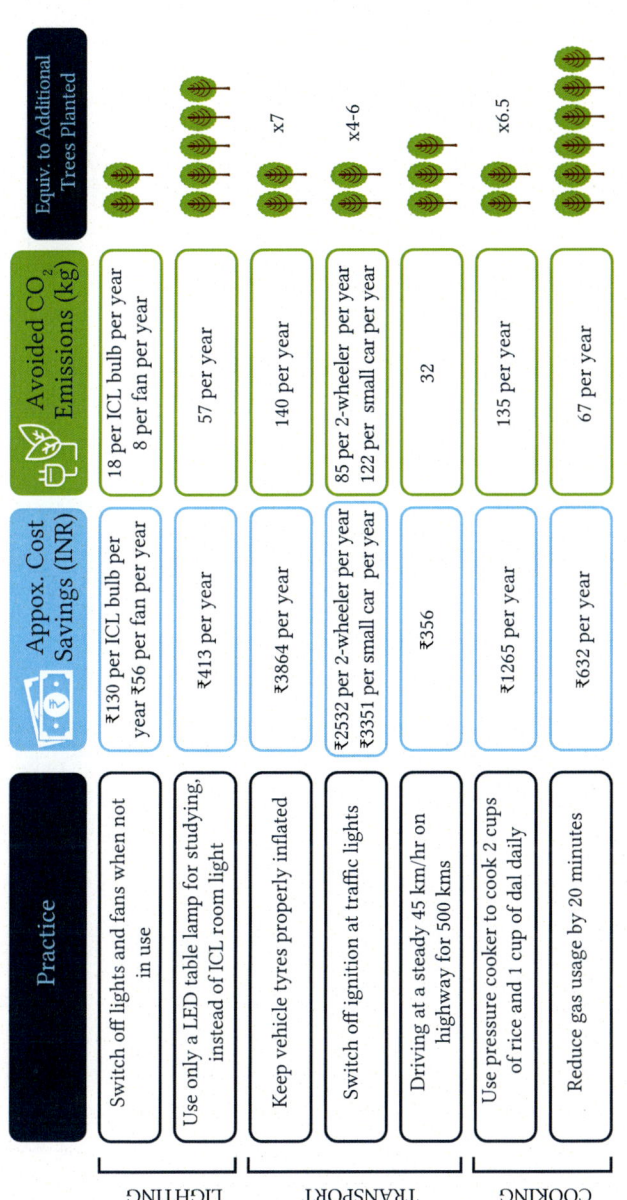

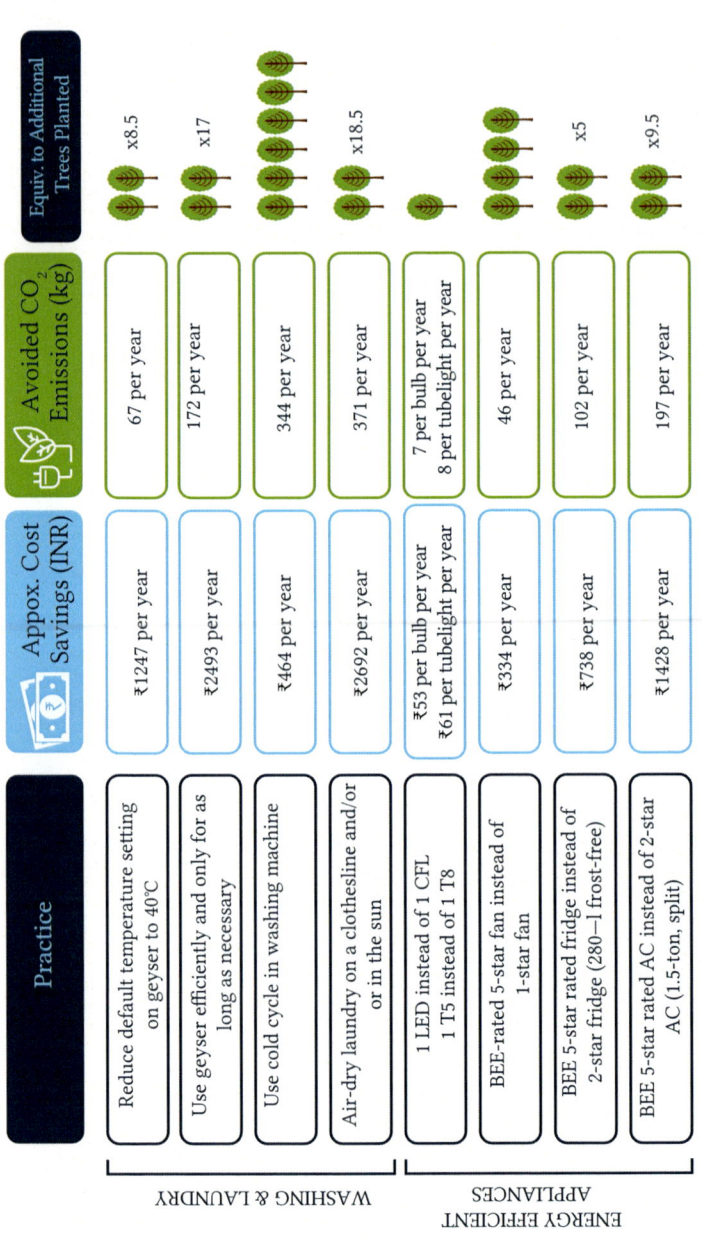

Figure 7 Potential impact of energy-efficiency practices and appliances

SOURCE: Ministry of Environment,Forests and Climate Change,2015(https://moef.gov.in/wp-content/uploads/2017/06/Low-Carbon-Lifestyles_0.pdf

When practised consistently and conscientiously energy efficiency can become a habit. Energy efficiency even at home can generate the change in mindset necessary to drive the actions of institutions that can have substantial impact in mitigating the adverse effects of climate change and safeguarding the future and quality of life of the next generation.

Choosing and building energy-efficient homes

Beyond the equipment and appliances used within, a home's interior design and external features can also significantly impact electricity consumption and the ways in which consumers engage with energy. These extend from the architectural design of the structure, the building materials, and insulation, if any, to the internal fixtures and furnishings. (The topic is covered in greater detail in *Green Homes and Workplaces* by Mili Majumdar and Minni Sastry in this series, namely 'Books for the Concerned Citizen'.)

Exterior

When selecting or building a home, consumers must consider its location and positioning, especially with respect to its surroundings, in addition to features that address the preferences and standard of living of its occupants. Areas in which homes are spatially distributed and shaded or surrounded by trees, other forms of greenery, or planned landscaping, are likely to be cooler in the summer months, thereby naturally reducing cooling needs.

Residents and home owners should also consider passive solar design, a key element of which is the (purposeful) orientation of the house to the sun's daily trajectory. Spaces with windows facing south are likely to receive uniform sunlight throughout the day; those facing west are likely to be the warmest; and rooms facing north and east will be relatively cool. Room layouts and space usage may accordingly be planned to minimize cooling in summer and heating in the winter. Areas that produce and radiate heat, like the kitchen and utility rooms, should face north or east, whereas spaces facing south may be preferred for living rooms, children's play areas indoors, and home offices to leverage natural light throughout the day.

The sun's path may also be considered when planning external shading, balconies, and patios as well as for determining the locations of walls and windows to address varying thermal retention and ventilation priorities during different seasons. Horizontal, vertical, or egg-crate (a combination of horizontal and vertical elements) overhangs can moderate the exposure of rooms to direct sunlight and the extent to which they remain warm during different times of the day whereas lattice walls can funnel air into and across spaces. While south-facing verandas may require shades to reduce the heat entering adjacent rooms, north-facing balconies and windows can help circulate air throughout the house, especially if openings and external vents are strategically located at different levels or heights and across rooms to create indoor wind tunnels and naturally facilitate air flow, thereby reducing dependence on heating or cooling devices.

Yet another element of passive design concerning the **building envelope** – the external walls, roof, and façade – is insulation. Insulated walls and roofs can moderate internal temperatures of a building significantly by trapping heat in winter and preventing cool air from escaping during summer. Insulation ranges from the materials used to build the walls and roofs – combinations of bricks and cement-plus-sand plaster, blocks of mineral wool, concrete, foam, or fibre rolls, for example – to pockets and cavities that trap air, to installing commercially viable products such as fibreglass and double- or triple-glazed panes for windows and doors. Reflective surfaces can also reduce the effect of solar radiation and can be achieved by painting outer walls and the roof with light-coloured reflective paint or plastering ceramic tiles.

Furthermore, growing plants on rooftops or building a rooftop garden as an additional buffer to absorb heat may also be considered while designing a building, a measure that also serves as a carbon sink. External landscaping and planting trees and bushes along the perimeter and external walls can provide a similar effect as well as acting as natural shading for outdoor electrical equipment and as a windbreak.

That said, the most critical element of insulating homes is to ensure that all outer walls, roofs, ceilings, doors, and windows are properly and completely sealed to prevent thermal losses. These seals and any cracks in the ceiling and roof should be checked at least annually, if not with the change of season, for seepage and leakage because walls and building structures respond to changes in external temperature and to weather events, undergoing natural wear and tear throughout the life of a structure.

Interior
Besides the external and structural features of a house, consumers can also adapt living spaces to make them more amenable to energy efficiency.

To begin with, in addition to the aforementioned insulating effort, customers may want to install light-weave curtains or blinds that assist ventilation during summer, while also blocking solar radiation that can heat spaces and increase reliance on electrical cooling equipment and appliances. During extreme cold conditions, these draperies may be replaced or reinforced with additional layers of heavier and thicker fabric to trap heat within the rooms. In assessing the aesthetic configuration of rooms, consumers are thus advised to also evaluate the colours and textures of such furnishings for their functional attributes.

Customers can also achieve similar effects by installing exterior shutters or awning or by covering window panes with clear or reflective film to reduce the glare and heat entering the home.

Flooring also plays an important role in insulating rooms and spaces. Marble and natural-stone flooring disperses heat effectively, keeping spaces cool for long periods, thereby reducing cooling needs during summer. On the other hand, wooden or parquet flooring can help to maintain a consistent ambient temperature by absorbing and retaining heat; combined with underfloor heating technology, such flooring can be especially effective in reducing reliance on space heaters and radiators during winter.

When installing new flooring, consumers may also consider adding insulating materials and layers between the ground and the waterproofing layer. The options include fillings or blocks of sand, clay, or straw, or even wooden beams and floor racks, as well as air and vapour barriers. In order to be effective, such floor insulation must be installed uniformly across the ground floor of the house and on upper floors of the house, as appropriate and necessary.

Although entailing upfront costs, new or replacement flooring requires little maintenance beyond daily cleaning, especially if treated appropriately. Customers can also always use carpets and dhurries, both to improve the aesthetic appeal of a space and to retain or dissipate heat, although in a much more localized and area-specific manner. The carpets and dhurries will need to be regularly vacuumed and washed to safeguard the respiratory health of occupants, and for overall safety and hygiene in the residence.

Homeowners and residents may also consider painting the walls in frequently used and naturally sunlit areas in light palettes or reflective colours to enable uniform distribution of light, and thus reduce the need for ceiling lights or floor lamps during daytime. Lighter-coloured walls also have the added advantage of making spaces appear warmer, and require less maintenance, because they display fading and wear less obviously than darker-coloured spaces do.

Finally, just as greenery outdoors can help reduce hot air entering the house, indoor plants can also contribute to ventilation and air purification, helping to establish a cool internal environment while releasing oxygen and absorbing harmful carbon dioxide. Plants with high transpiration rates like palms and different varieties of Ficus are especially known for improving humidity and air quality in closed spaces, and require little maintenance and sunlight.

Equipment and energy audits

As mentioned earlier, appliances and devices used in a home can significantly impact its electricity and energy consumption. This holds true, on occasion disproportionately, for even large items of equipment used for making a house comfortable and habitable.

While heating and cooling constitute a significant portion of a household's electricity costs, residents and owners of larger homes may consider investing in energy-efficient HVAC systems that centralize ventilation and maintain ambient temperatures throughout homes with packaged air-conditioner units. These systems are also programmable and when combined with appropriately positioned air ducts that distribute air throughout the house, reduce not only energy losses but also GHG emissions. Ductless split cool ACs and heat pump systems may also prove to be energy-efficient options for households looking for complete temperature-controlled domestic air and water solutions.

Consumers may want to install exhaust fans in the parts of the house that either generate or trap heat, such as kitchens and bathrooms, and closed areas that store large equipment, to reduce the home's overall internal temperature during summer. The exhaust fans will also prevent the growth of mildew and harmful bacteria that thrive in closed and humid environments.

Consumers may supplement electricity supply by installing solar photovoltaic (PV) panels on rooftops or in open, shade-free spaces, either factoring in these systems while designing a home or retrofitting existing homes. (For more information, see *Sun Through the Roof* by Suneel Deambi and Shirish S Garud in this series, namely 'Books for the Concerned Citizen'.) With its potential to reduce dependence on the grid, transitioning to, or incorporating, solar power will significantly reduce the consumer's carbon footprint, while also lowering dependence on inverters and external generators during power shortages and, when deployed in combination with battery storage technology, even during periods of little or no sunlight. Along with the incentives offered by the central government and state governments for adopting solar energy, reduced lifetime energy costs can significantly offset upfront investment in solar panels, which have also become increasingly affordable due to ongoing transformations in the domestic market.

Consumers may routinely consult and engage certified professionals to conduct home energy audits to check the effectiveness of electricity supply and consumption in homes, and to identify any inherent inefficiencies or potential risks of malfunction in internal circuits. Albeit a short, but necessary, exercise, energy audits offer a detailed assessment of how a household is using electricity based on physical checks of the circuits, equipment and devices being used, an evaluation of the effectiveness of insulation and ventilation systems, and identification of possible electricity theft and potential leaks or cracks in the walls or ceiling. Energy audits thus guide users on how they can reduce electricity consumption and incorporate energy efficiency into their daily activities, especially highlighting how consumers can eliminate inefficient use of electricity from devices left on in standby mode.

The energy auditor or an electrician may also recommend metering different circuits within the house separately, known as sub-metering, to monitor energy consumption more effectively across components, and to ensure that larger equipment and appliances that consume proportionately more electricity than the rest of the household are operating efficiently. Sub-metering is also relevant if multiple nuclear family units are occupying different parts of a structure served by a single meter, and want more flexibility and autonomy in their electricity consumption.

Although several accredited independent companies and agencies offer energy auditing services, the electricity or power departments and discoms also train and empanel professionals as home energy auditors. Consumers are advised to consult their discom, as well as building managers, about local and experienced agencies and professionals who can conduct walk-through, standard, or detailed audits, as applicable and appropriate.

Building codes and regulations

The above considerations are also covered in several documents prepared by different government agencies and related institutions to assist consumers in evaluating and building homes for energy efficiency. In 2018, the Union Ministry of Power and Bureau of Energy Efficiency released *Eco-Niwas samhita 2018 (Energy Conservation Building Code for Residential Buildings)*, a code to guide architects, developers, builders, and home owners on active and passive strategies for developing building envelopes and systems that leverage natural elements to address lighting, insulation, and ventilation requirements of consumers while also incorporating the principles of energy efficiency and environmental protection. Along with requirements and recommendations for energy-efficient electrical and mechanical equipment used in building operations and renewable energy generation, the code establishes minimum performance standards for new building projects occupying plot areas greater than 500 square metres and is intended for incorporation by urban local bodies into building by-laws. Details about the code can be accessed on econiwas.com. The above code is supplemented with a web-based guidance manual, *The Handbook of Replicable Designs for Energy Efficient Residential Buildings*, which guides building designers, owners, and architects on replicable designs for energy-efficient homes.

Furthermore, the central government also announced a scheme to assign energy-efficiency labels to residential buildings in 2019. With an associated online star-rating tool, the programme enables consumers to transparently assess and benchmark the energy performance of a house, thereby encouraging buyers and sellers to include energy-efficiency considerations into their pricing and purchasing decisions, and to identify opportunities for reducing energy consumption in homes.

These efforts are supplemented by GRIHA (green rating for integrated habitat assessment), an initiative of TERI (The Energy and Resources Institute). The initiative is a government-endorsed system for rating newly constructed green buildings on the basis of building performance and design on several metrics, including energy efficiency and ecological sustainability, throughout the lifecycle of the building, starting from planning and construction through occupancy and maintenance.

Increasingly, state governments across the country are notifying and enforcing these codes and rating systems for new building constructions. Authorities and municipal bodies across the country have also endeavoured to incentivize construction of energy-efficient buildings and residences that incorporate elements of ecological preservation and reduced consumption of fossil-based energy, including adoption and retrofitting of new and existing structures with solar-powered equipment.

Thus, in addition to professional energy audits before purchasing or renting homes, especially those less than 5 years old, consumers are also advised to seek clarity and information from real-estate agents and developers on the compliance of residential buildings with energy-efficient building mandates and guidelines, as well as on the recognition of buildings and residential complexes by reputable national and international awards programmes for green residential buildings. Residents and home owners are also advised to consult certified green building development consulting and audit agencies for guidance on the building materials used for construction, as well as on the appliances and equipment installed in the home to maximize utility and energy efficiency throughout the life of the equipment.

Because of the growing popularity of zero-carbon homes across the globe, as well as the more intensely regulated nature of the residential real-estate sector, a wealth of resources has been developed, and are available for home owners and residents interested in building low-consumption and energy-efficient homes. Most of these guides are developed keeping in mind the varying geographies and environments, and are developed by stakeholders like regulators, civil engineers, architects, and real-estate industry with deep knowledge of building green and sustainable homes. For a vast country like India, whose residents experience a variety of climate zones and weather patterns due to geographic dispersion, such domestic and international resources are particularly relevant.

The information given above aims to be a primer on such resources, highlighting considerations to make living spaces energy minimal and guiding consumers on energy-efficient design

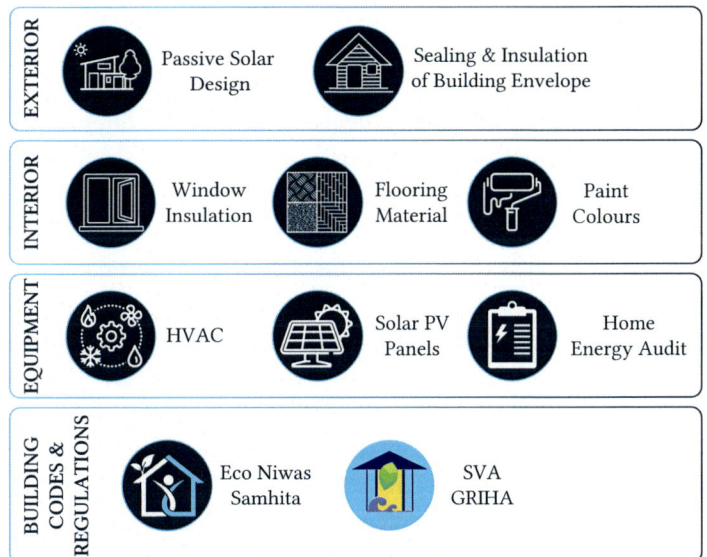

Figure **8** Considerations for energy-efficient home design

elements that could be incorporated in their building designs or prospective homes. These suggestions are applicable to all types of habitation including stand-alone bungalows and apartments in multistorey buildings. Consumers are strongly advised to consult the larger universe of professionals, books, manuals, and guidelines that elaborate on each of the aforementioned elements (Figure 8) in greater detail for a more comprehensive and holistic understanding of selecting or building homes that are sustainable, locally-relevant, and have the least adverse impact on the environment.

Energy efficiency beyond the home

Measurable progress towards climate resilience and energy security can be achieved when consumers practise energy efficiency not only inside their homes, but also outside. At an institutional level, large-scale adoption of energy efficiency in commercial and industrial sectors, which claim a significant share of energy consumption both in India and abroad, is yet to reach its potential, partly because the technology is constantly evolving and developing, but also because the rules for sustainable climate- and energy-conscious action are yet to translate into organizations' operating and capital allocation decisions. Greater awareness and understanding among leaders, decision-makers, and employees, and the subsequent change in mindset on integrating energy efficiency into the operations and offerings of private- and public-sector players can have significant effects, and enable organizations to be purpose-led and impactful in their respective journeys towards environmental and business sustainability.

The range of possible interventions thus covers both individual and institutional actions. At the individual level, employees could ensure that they do not leave machines or equipment idling, switching them off when not in use and at the end of the workday, either carpooling or using public transport to commute between the home and workplace, conscientiously switching off lights and ventilation devices when exiting rooms, and taking the stairs instead of elevators or escalators to access different floors as much as possible.

To reduce emissions and pollution, employees may also conduct meetings and conferences virtually instead of in-person, reuse partially used paper, and carry refillable bottles and cups and reusable cutlery to their offices to reduce energy used in recycling plastics.

Organizations play an important role in facilitating such resource-conscientious actions at the individual level. Along with ensuring that energy efficiency is a core component of their respective environment and sustainability policies, institutions may consider retrofitting office spaces with energy-efficient lights, motors, pumps, and HVAC systems; installing occupancy sensors in conference rooms and passageways; and, programming computer stations and laptops to switch to sleep mode if left idle for a stipulated duration. Several institutions also offer shuttle or bus services to assist employees' daily commute, thereby also reducing the size of the carbon footprint due to transport.

Organizations may also consider locating their offices in buildings that have installed, or are willing to install, solar panels to supplement the buildings' energy supply, as well as charging stations in open-air garages for electric vehicles. Discoms across the country, especially those serving special economic zones (SEZs) and large commercial complexes, may partner with companies willing to install wind turbines and solar panels on their premises to generate energy that can be supplied to the grid, thus reducing the organizations' electricity costs.

The above practices are age- and geography-agnostic, and can very well be applied in a variety of contexts, including schools and educational institutions. Given the urgency of action, the principles of resource efficiency must be instilled at a young age and permeate the entire span of India's age and socio-economic demographics. In 1991, BEE established NECA (National Energy Conservation Awards) for industries, and since 2003, has been conducting an annual national painting competition in which schoolchildren and youth from across the country engage in issues relating to environment conservation, and, through art, express their understanding of energy efficiency and the impact of human activity on the planet. The winners are felicitated at an annual awards ceremony that also recognizes organizations across industrial sectors such as power, iron, steel, buildings, and transport and manufacturers of energy-efficient appliances for their contributions towards universal adoption of energy efficiency, including reducing their own energy consumption and carbon emissions.

Such recognition through the awards is a small component of the much broader efforts by the central government and BEE under NMEEE (National Mission for Enhanced Energy Efficiency) to engage large- and medium-scale industry in dialogue and action on energy efficiency. Initiated under India's National Action Plan on Climate Change in 2008, NMEEE anchors the regulations, policies, and initiatives for driving adoption of energy-efficient practices by industry and consumers, as well as the transformation and acceleration of the market for energy-efficient electrical equipment and appliances.

The national mission is executed primarily through four schemes under the auspices of the Ministry of Power and BEE.

PAT, or perform, achieve, and trade

Aimed at helping individual industries reduce their specific energy consumption in energy-intensive sectors like cement, iron, steel, paper, pulp, and textile and associated industries, including railways and refineries, the PAT scheme benchmarks the industries' energy consumption, establishes individual targets, and measures annual incremental improvements in energy efficiency. Industries of varying sizes can participate in the scheme, ranging from MSME (micro, small, and medium enterprise) clusters to large industrial complexes. Industries that exceed their respective energy savings targets become eligible to convert the excess savings into energy savings certificates, which can be traded at power exchanges. The PAT scheme has enlisted more than a thousand industries in 13 sectors across the country since its inception in 2012, with newer industries and sectors added to its ambit every year. By March 2023, the industries participating in the scheme will have reduced annual carbon emissions by 110 million tonnes and achieved 26 MTOE in avoided capacity generation (Bureau of Energy Efficiency 2021).

MTEE, or market transformation for energy efficiency
The scheme supports demand aggregation programmes by discoms and energy-service companies such as the Energy Efficiency Services Ltd to increase the adoption of energy-efficient appliances, making them affordable and accessible. Such demand aggregation programmes have transformed the domestic markets for LED bulbs and fluorescent tubes, which have seen significant nationwide adoption, consequently reducing their market prices by more than 70% since 2015 while also expanding the number of domestic industry players for this product category, and for super-efficient air conditioners. As part of these market transformation efforts, EESL has retrofitted nearly 10 000 buildings across the country with energy-efficient equipment and appliances, with subsequent energy audits revealing savings of 30% to 50% in these buildings' energy costs.

Similar demand aggregation-based market transformation initiatives have also been successfully undertaken and are ongoing for solar-powered water pumps for agriculture, solar-powered LED street lamps, smart meters, and electric vehicles and EV charging stations. Such interventions continue to lower annual carbon emissions by more than 40 million tonnes, and nationwide energy consumption by billions of units of energy every year.

EEFP, or the energy efficiency financing platform
To strengthen financing to promote energy-efficiency adoption and adaptation by industries, the EEFP raises awareness and provides capacity-building services to financial institutions to enable them to evaluate the merits of, and supply funding for, energy-efficiency programmes undertaken by individual industries. To this effect, BEE hosts a series of platforms to facilitate engagement and understanding between financial institutions, manufacturers of energy-efficient equipment, and organizations looking for capital to help them adopt energy-efficient technologies, thus demonstrating the business impact and the investment and return potential of energy efficiency at industrial levels. The series, 'Investment bazaar for energy efficiency', held in partnership with state power and energy departments and municipal bodies across the country is amongst the most prominent initiatives under this scheme.

FEEED, or the framework for energy efficient economic development
Extending on EEFP, the FEEED scheme oversees the design and development of financial instruments to facilitate funding of energy-efficiency initiatives of organizations and industries. The partial risk-sharing facility established under this scheme in partnership with the Clean Technology Fund and the Global Environment Fund supports financial institutions with partial credit guarantees on loans they extend to industry for energy-efficiency projects, while the energy-efficiency financing facility refinances banks that have lent money to large industries and MSME clusters for investments in upcoming technologies like smart grids and EVs.

In addition to the above schemes, BEE also conducts capacity-building programmes for industries and MSME clusters to help them identify more holistic approaches to energy efficiency, even preparing cluster-specific manuals and detailed project reports, and conducting training programmes on energy-efficient technologies.

Furthermore, BEE also developed the Energy Conservation Building Code in 2017, which established minimum standards of energy consumption and efficiency for new commercial buildings through a design-led approach. Applicable nationwide, the code considers different climatic zones across the country in prescribing standards for the building envelope, indoor and outdoor lighting, HVAC systems, solar-powered water-heating systems, as well as electrical infrastructure for buildings. Similar to the Eco-Niwas guidelines, the code identifies options for building designers and developers to integrate renewable-energy-based consumption and passive design strategies.

Although developed at the central level, the authority to adopt, adapt, and enforce the code lies with state governments, individual municipalities, and urban bodies. Central bodies such as BEE support state and city entities by identifying third-party certifiers and demonstrating the impact and effectiveness of the code on select project sites.

In each aspect, whether inside their homes or outside, consumers should assess the effectiveness of their choices and efforts to be energy efficient, comparing resource input with the performance output of the equipment they use and the buildings they occupy, and benchmark these measures before making new investments. Similar to the energy consumption and efficiency metrics stated on the BEE star labels on consumer appliances, and fuel efficiency of cars, colloquially known as 'mileage', the metric calculated for commercial buildings is called the Energy Performance Index, which compares the total energy consumed by a building in a year to its total built-up area: the lower the index, the more energy efficient the building.

Before occupying or investing in buildings, whether as short-term or long-term leaseholders, decision-makers are encouraged to review the energy performance index and also consider certification by reputable independent third-party verifiers, such as the CII, India Green Building Council, and Green Business Certification Inc, which review the impact of new constructions from a holistic ecological perspective.

Thus, even in their professional lives, consumers can avail themselves of a vast array of opportunities and initiatives that enable them to be energy efficient, whether as employees using available resources in offices conscientiously, as administrators and building managers assessing and adapting buildings for sustainability and energy savings, or as leaders endeavouring to improve cost and operational efficiencies of their respective businesses while mitigating adverse impact on the environment and surrounding communities. Both in their professional capacity as stakeholders and decision-makers in industry, and in their individual capacity as members of a household, consumers are encouraged to keep themselves abreast of such initiatives, and the incentives and support offered by the government and the industry to promote widespread adoption of energy efficiency.

Frequently asked questions

In this book, we have tried to educate electricity and energy consumers about the principles, benefits, and practices of adopting energy-efficiency measures. However, we recognize that the concerned citizen may still have questions. Therefore, in this section we address the most frequently voiced concerns about energy efficiency and related equipment.

Q. 1. How will my individual actions of switching off appliances and buying energy-efficient devices really help the nation or the planet?

Each individual's efforts towards energy efficiency and conscientious use of resources can reduce their respective household's electricity consumption by over 50%. For each hour that an unused 60-watt light remains switched off, at least 450 grams of coal can be saved from burning to generate electricity, translating to an equivalent weight in avoided carbon emissions.

The cumulative impact of such energy efficiency measures for all the light fixtures and appliances in households across the country could be monumental, equivalent to emissions reduction of 14 million tonnes of CO_2 and 12% of the nation's potential for energy savings. For a start, this could mean lesser air pollution and cleaner air in urban areas, as well as uninterrupted electricity supply of better quality.

As a whole, energy efficiency could reduce India's energy intensity by 19% by 2040 and contribute up to 56% to India's 2030 target for reduced emissions intensity, with the residential sector forming a significant part of the country's and the world's ambitions for mitigating climate change. The importance of each household, and the actions of each of its members, to the worldwide effort to avoid a global temperature rise thus cannot be emphasized enough, with every action having multifold repercussions.

Q. 2. Why should energy efficiency be a priority for a developing country like India when we have so many other competing development concerns?

India's economy needs to grow significantly for the country to address the multiple development concerns and ambitions of its rapidly growing population, the second largest in the world. This economic growth will be accompanied by a parallel, if not accelerated, pace of energy consumption to drive industrialization, urbanization, infrastructure development, and rising incomes commensurate with the country's development priorities. Such growth in energy demand and supply is essential for development, and to the goal of bringing modern energy services to millions of Indians who lack such access today.

The mission for the requisite universal, meaningful, reliable, uninterrupted, and affordable energy access will be greatly assisted by energy efficiency, along with the imperative to address concerns related to quality of life and growing emissions intensity. The International Energy Agency estimates that more than 55% of the options to reduce emissions of GHGs relate to energy efficiency and have the added benefit of reducing the burden of energy costs, a growing concern in today's increasingly resource-constrained conditions.

Yesterday's resource-intensive practices will not sustain the India of today or tomorrow. Energy efficiency is thus at the very centre of India's efforts to address all of its other development priorities, including and especially, just climate action.

Q. 3. Will practicing and adopting energy efficiency compromise my comfort and convenience, or prove to deliver rewards corresponding with the effort? Why aren't the government and public and private sectors doing more to make infrastructure and industry energy efficient, and therefore reduce this imposition on households and individuals?

Collectively, the government and public and private sectors have undertaken several initiatives in the industrial and commercial sectors, as well as initiated measures to make energy efficiency convenient and relevant for household and residential consumers. These efforts have resulted in electricity savings of nearly 10% in 2018/19 alone (Press Information Bureau 2020), and the industry and government will continue to seek opportunities to deploy more measures across sectors and stakeholder groups.

That said, consumption by India's residential sector, comprising primarily of households and individual consumers, constitutes nearly a quarter of the country's energy demand, and by extension, a corresponding portion of carbon emissions. This proportion is by no means insignificant, and if efforts by the government and the industry towards energy transition, universal energy access, energy security, and climate change mitigation are to succeed in achieving sustained and meaningful impact, collaboration and contribution of the residential sector across the country's socio-economic demographics are critical.

It is also important to note that nearly 70% of India's population is of working age, and therefore inasmuch as these people are members of their respective households, a significant proportion of them are also part of the industry and government bodies that are endeavouring to develop solutions that will have a meaningful impact on climate action and strengthen India's energy security. Adoption of energy efficiency by this group of people, and their approach to resource conservation in their homes and personal lives, thus also have significant repercussions on their approach to climate change at the national level.

Finally, it is important to note that consumers are direct beneficiaries of the energy-efficiency practices and technologies they adopt, which can lower their electricity bills by up to 20%, often within one billing cycle, and, in turn, enable them to redirect the accrued savings to other aspects of their lives for their comfort, convenience, and, in many cases, their earning potential. Households and individual consumers are thus critical stakeholders in realizing India's ambitions related to sustainability, development, and climate action.

Q. 4. Are energy-efficient devices expensive? If so, why should I buy an expensive device when I will change it in three years?

The total cost of any appliance comprises its acquisition cost (typically 20% or less than the cost of owning the appliance over its lifetime), its operating cost, and the cost of energy for operating the appliance. The savings due to lower lifetime energy consumption frequently exceed the difference in the prices of efficient and less efficient appliances at the time of purchase.

Consumers may also appreciate that, due to ongoing transformation of the industry and progress in technology, energy-efficient models of many appliances and devices are now price-competitive with their less efficient counterparts, and are also readily and widely available in the market, LED bulbs and fluorescence tubes being the most evident and prominent examples.

That said, even for devices where such price parity is yet to be achieved, the life of energy-efficient appliances is typically longer than that of conventional devices, with the maintenance and lifetime costs of the former also being lower. Furthermore, the electricity savings that accrue to consumers from switching to energy-efficient appliances can help to significantly offset their upfront investment costs, with the payback period ranging from three to five years for most appliances currently available in the market.

With increasing availability of technologies that enhance the energy efficiency of new products, we can expect the payback period to shorten in the coming years, making energy-efficient models more affordable to consumers across multiple income groups throughout India.

Q. 5. I will invest in an energy-efficient device; will others?

Prevalent sociological studies and research indicate that consumers are most likely to emulate energy-conscious behaviour and replicate the buying choices of their friends, family members, neighbours, and other people they know and trust. Thus, inasmuch as consumers are responsible for the individual choices they exercise in adopting energy efficiency, they are also greatly influenced by the buying decisions of others. Although the government and the industry constantly try to raise consumer awareness about, and increase action towards, energy efficiency adoption, the onus also lies on citizens, such as the readers of this book, to discuss their experiences of energy-efficient devices and practices, and the benefits they have enjoyed, with people they know and who seek their advice, and to participate in efforts to educate their own and surrounding communities. Collective alignment and action on the imperative for resource-conscientiousness can have far-ranging impact and contribute to slowing down climate change and sustaining environmental health for the present and future generations.

Q. 6. When should I replace the electronic devices or equipment in my home with their more energy-efficient models?

Consumers may buy energy-efficient appliances either when their current appliances need to be supplemented or replaced to meet their current and future needs, or when the output from the existing appliances is less than their corresponding electricity input or declining, especially after repairs. In either case, consumers may first evaluate and consider BEE-star-labelled appliances, as even the appliances for which the labels are not mandatory are routinely spot-tested to ensure that they meet stated efficiency and delivery metrics.

Consumers must also explore exchange options with trusted retailers when replacing appliances to avail themselves of the opportunities to offset the upfront costs of new equipment and ensure that the old appliances are either being recycled or being disposed of responsibly, thereby also reducing e-waste that can leak harmful toxins into the environment, eventually affecting human and ecological health adversely.

Consumers must also routinely check the energy efficiency and star ratings of products available in the market at any given time, as efficiency standards are periodically revised and minimum efficiency benchmarks raised. Thus, a product that carries a 5-star rating today may become a 4-star appliance two years from now as new and more efficient technology enter the market. Exchanges and routine upgrades to increasingly efficient equipment also ensure that the appliances being sold in secondary markets are also efficient, subsequently also reducing the overall load on the grid and facilitating universal and uninterrupted supply of electricity.

Q. 7. Why is my energy-efficient device not delivering the promised savings?

Consumers should be aware of three major facts when using electrical appliances, equipment, and devices, regardless of whether they are labelled energy efficient or not.

First, energy-efficient devices can deliver savings when used with the same intensity as their less efficient counterparts. Consumers may therefore review the manner in which, and the duration for which, they use energy-efficient devices at optimal levels.

Second, a technician, professional, or equipment manufacturer should be consulted on the functioning of the device to its stated performance standards, and accordingly consumers should seek any repairs or servicing to address any malfunction.

Third, an electrician must be consulted to ensure that the connection has not been compromised. Consumers may also want to request home energy audits to identify any potential leaks, losses, or unknown electricity usage of the sanctioned load or to rectify vulnerabilities in the residence's insulation or sealing mechanisms.

Q. 8. The country is gradually transitioning to renewable energy, which costs almost equivalent to, or even less than, thermal energy or fossil fuels. Is energy efficiency still necessary and relevant?

Yes, energy efficiency continues to be relevant and necessary, both from an individual standpoint, and in the national and global interests. Regardless of the source of energy, energy-efficient practices alone can help households reduce their electricity costs by up to 20%, with the benefits being felt within one billing cycle. The subsequent load reduction on the grid can thus not only facilitate more consistent and universal electricity access across the country, thereby enabling greater socio-economic development and mobility, but also strengthen India's energy security, reducing our nation's dependence on energy and oil imports for sustaining our economy and livelihoods.

Energy security assumes greater significance in the context of the transition to renewable energy, because with increased capacity to generate electricity from renewable energy, universal adoption of energy efficiency, and moderation in domestic consumption, India could very well change from being an energy importer to an energy exporter, thus strengthening the country's macro fundamentals and positioning in the global economy.

That said, from a broader perspective, electricity available from even renewable sources is volatile because it is subject to weather conditions and will continue to be in limited supply until technology evolves sufficiently to enable countries and industries to achieve capacities in capturing, converting, storing, transmitting, and distributing renewable energy in a commercially viable manner. Despite progress in the ongoing global energy transition, the world as a whole, and emerging and developing markets such as India in particular, are not going to be able to transition fully to renewable energy in the short to medium term, entailing continued dependence on fossil fuels and the accompanying emissions. A key objective of sustained energy efficiency is to curtail the need for fossil-fuel-based energy generation, thus reducing the harmful emissions that accelerate the threat of climate change and deepen its impact on the ecological balance of the planet and the livelihoods of its inhabitants.

Conclusion

The need, intervention potential, and possible impact of energy efficiency are unquestionable. What this book has tried to provide household energy consumers is greater awareness of energy efficiency, assistance in translating this understanding into action, as well as a deeper appreciation of the ease and convenience with which energy efficiency can be incorporated into everyday life.

To this end, we hope that the contents of this book will help its readers either to embark on or advance in their journey towards energy efficiency, and will also encourage them to seek additional guides and resources to continue educating their communities and themselves to mobilize a national movement towards resource efficiency and conservation. Only through holistic approaches to resource use can systemic changes to mitigate the adverse effects of climate change be achieved and have the desired impact.

While refining their understanding of energy efficiency and climate change, consumers are encouraged to consult the wealth of resources offered by many institutions and organizations like BEE, TERI, IEA, United Nations Environment Programme, The World Bank, World Resources Institute, Council on Energy, Environment and Water, Alliance for Energy Efficient Economy, Centre for Study of Science, Technology and Policy, Centre for Science and Environment, and countless others that advance the cause of climate action with truly ground-breaking work in policy advisory and design, as well as with frameworks, partnerships, and community outreach initiatives towards making energy efficiency applicable and relevant to stakeholders across the value chain.

Consumers are also encouraged to consult the resources developed by the discoms in their respective jurisdictions to guide them on efficient resource management and use. An admirable nationwide initiative promoted by several discoms has been the inclusion of tips on energy efficiency in consumers' electricity bills. A compendium of similar such initiatives is available on BEE's website (https://beeindia.gov.in/).

An especially critical component of meaningful impact is the commitment of efforts and resources towards facilitating discourse and dialogue amongst consumers on their experiences with adopting energy efficiency. Although significant efforts are under way to highlight government and industry initiatives, creating similar visibility and recognition opportunities to highlight the agency and impact by consumers towards energy efficiency can facilitate a ripple effect throughout communities across the country.

Finally, inasmuch as this book has been developed for consumers and audiences in India, the development process has nonetheless entailed consultation of resources and best practices from across the world, in itself speaking to the universal applicability of energy efficiency to global climate and sustainability movements. To this effect, we have every hope that international audiences will also find the principles and guidelines outlined in this book to be relevant, localizing and applying them to their respective environments, living conditions, and quality-of-life priorities.

References

Agrawal S, Mani S, Aggarwal D, Kumar C H, Ganesan K, Jain A. 2020. *Awareness and adoption of energy efficiency in Indian homes: insights from the India Residential Energy Survey (IRES) 2020*. New Delhi: Council on Energy, Environment and Water. 60 pp. https://www.ceew.in/sites/default/files/CEEW-IRES-Awareness-and-adoption-of-EE-in-Indian-homes-07Oct20.pdf

Ali S. 2018. *The future of Indian electricity demand: how much, by whom, and under what conditions?* Brookings India. 50 pp. https://www.brookings.edu/wp-content/uploads/2018/10/The-future-of-Indian-electricity-demand.pdf

Alliance for Energy Efficient Economy. 2021. *India's energy efficiency landscape: a compilation of policies, priorities and potential*. New Delhi: AEEE. 78 pp. https://aeee.in/wp-content/uploads/2021/05/India%27s-Energy-Efficiency-Landscape-Report.pdf

Bureau of Energy Efficiency. 2021. *Annual Report 2020-21*. New Delhi: BEE. 120 pp. https://beeindia.gov.in/sites/default/files/BEE%20Annual%20Report%20-%20%28English%20Final%29.pdf

Central Electricity Authority. 2020. All India electricity statistics: general review 2020 (containing data for the year 2018-2019). New Delhi: CEA. 250 pp. https://cea.nic.in/wp-content/uploads/general/2020/GR_2020.pdf

International Energy Agency. 2015. Energy efficiency outlook for India: sizing up the opportunity. Paris: IEA Publications. 72 pp. https://iea.blob.core.windows.net/assets/a27eeef7-b4ef-44f2-b7ca-0c85db33d0b7/IndiaEnergyEfficiencyOutlook_20161215_Final.pdf

International Energy Agency. 2021. *India Energy Outlook 2021*. Paris: IEA Publications. 250 pp. https://iea.blob.core.windows.net/assets/1de6d91e-e23f-4e02-b1fb-51fdd6283b22/India_Energy_Outlook_2021.pdf

Kajal K. 2022. Scorching weather forces India to face climate change head on. *Al Jazeera*, 2 May. https://www.aljazeera.com/news/2022/5/2/scorching-weather-forces-india-to-face-climate-change-head-on

Krishnan M. 2022. Climate change: Why it is now or never for India. *The Telegraph Online*, 3 March. https://www.telegraphindia.com/india/climate-change-why-it-is-now-or-never-for-india/cid/1854355

Mathur R, Shekhar S, Sethi G, Kumar S. 2018. Energy efficiency & India: the possibilities ahead. The Energy and Resources Institute. https://www.teriin.org/article/energy-efficiency-india-possibilities-ahead

Ministry of Environment, Forest and Climate Change. 2015. *Low Carbon Lifestyles*. New Delhi: MOEFCC. 26 pp. https://moef.gov.in/wp-content/uploads/2017/06/Low-Carbon-Lifestyles_0.pdf

Ministry of Environment, Forest and Climate Change. 2021. India: Third Biennial Update Report to the United Nations Framework Convention on Climate Change. New Delhi: MOEFCC. 480 pp. https://unfccc.int/sites/default/files/resource/INDIA_%20BUR-3_20.02.2021_High.pdf

Ministry of External Affairs. 2021. National Statement by PM Modi at COP26 Summit in Glasgow. 1 Nov. https://www.narendramodi.in/national-statement-by-pm-modi-at-cop26-summit-in-glasgow-558205

Ministry of Power. 2021. Implementation of energy efficiency measures in India saves substantial amount of CO_2 emissions in the country. [Press release] https://pib.gov.in/PressReleasePage.aspx?PRID=1725448

Ministry of Power. 2022. Power sector at a glance all India. https://powermin.gov.in/en/content/power-sector-glance-all-india

Pimpalkhare A. 2020. Expanding India's energy efficiency sector. New Delhi: Observer Research Foundation. https://www.orfonline.org/expert-speak/expanding-indias-energy-efficiency-sector

Press Information Bureau. 2020. The energy efficiency initiatives by BEE leads to savings worth Rs. 89,122 Cr. in 2018-19. 6 May https://pib.gov.in/PressReleasePage.aspx?PRID=1621501

PTI. 2021. Assam, Andhra, Bihar, Karnataka, Maharashtra most vulnerable to adverse climate events, says study. *The Economic Times*. 26 Oct. https://economictimes.indiatimes.com/news/india/assam-andhra-bihar-karnataka-maharashtra-most-vulnerable-to-adverse-climate-events-says-study/articleshow/87287338.cms?utm_source=contentofinterest&utm_medium=text&utm_campaign=cppst

Saleem Q, Garg A, and Mendonca B. 2020. Opinion: power to the people - a case for energy-efficient appliances in rural India
https://energy.economictimes.indiatimes.com/news/power/opinion-power-to-the-people-a-case-for-energy-efficient-appliances-in-rural-india/77736653

Spencer T and Awasthy A. n.d. Analysing and projecting Indian electricity demand for 2030. New Delhi: The Energy and Resources Institute. 42 pp.
https://www.teriin.org/sites/default/files/2019-02/Analysing%20and%20Projecting%20Indian%20Electricity%20Demand%20to%202030.pdf

The World Bank. 2022. Access to electricity (% of population) - India.
https://data.worldbank.org/indicator/EG.ELC.ACCS.ZS?end=2020&locations=IN&start=1993&view=chart

World Meteorological Organization. 2021. State of the Climate in Asia 2020. 36 pp. [WMO-No. 1273]
https://library.wmo.int/doc_num.php?explnum_id=10867